Poultry Science

Poultry Science

Edith Powell

□SYRAWOOD
PUBLISHING HOUSE

New York

Published by Syrawood Publishing House,
750 Third Avenue, 9th Floor,
New York, NY 10017, USA
www.syrawoodpublishinghouse.com

Poultry Science
Edith Powell

International Standard Book Number: 978-1-64740-081-1 (Hardback)

Cataloging-in-Publication Data

Poultry science / Edith Powell.
    p. cm.
Includes bibliographical references and index.
ISBN 978-1-64740-081-1
1. Poultry. 2. Aviculture. 3. Eggs--Production. 4. Animal culture. I. Powell, Edith.
SF487 .P68 2022
636.5--dc23

Table of Contents

Preface

It is with great pleasure that I present this book. It has been carefully written after numerous discussions with my peers and other practitioners of the field. I would like to take this opportunity to thank my family and friends who have been extremely supporting at every step in my life.

The domesticated birds that are kept by humans for their feathers, meat or eggs are collectively referred to as poultry. Some of the birds which fall under this category are chickens, geese, ducks and turkeys. Poultry science is a specialization of animal science which is concerned with finding improved and more effective ways of raising poultry that produce healthy eggs and meat. It works to improve poultry production and food safety. It also studies the different diseases which can affect poultry production, their prevention and the ways in which they can be controlled. This book provides comprehensive insights into the field of poultry science. Different approaches, evaluations and concepts related to this field have been included herein. Those in search of information to further their knowledge will be greatly assisted by this book.

The chapters below are organized to facilitate a comprehensive understanding of the subject:

Chapter – Introduction

Poultry science is involved in the study of various types of birds which are raised for eggs, meat and feathers. It also studies the processes of incubation and hatching, as well as poultry products, poultry breeding and poultry processing. This chapter has been carefully written to provide an easy understanding of these facets of poultry science.

Chapter – Types of Poultry

There are numerous birds which fall under the category of poultry. Some of them are chicken, ostriches, goose, domestic ducks, domestic pigeon, squabs, peafowls, domestic turkeys and guinea fowls. The topics elaborated in this chapter will help in gaining a better perspective about these types of poultry.

Chapter – Poultry Farming

The form of animal husbandry which is involved in raising domesticated birds such as chickens, ducks, turkeys and geese to produce meat or eggs for food is termed as poultry farming. Some of the birds which are raised under poultry farming are ducks, broiler chickens and peacocks. The chapter closely examines the farming of these types of poultry to provide an extensive understanding of the subject.

Chapter – Poultry Management

The husbandry practices and production techniques that help to increase the efficiency of poultry production is referred to as poultry management. The main practices which fall under poultry management are brooding management, poultry housing and management, poultry rearing, chick culling, forced molting, etc. This chapter has been carefully written to provide an easy understanding of these methods and techniques of poultry management.

Chapter – Poultry Diseases and their Management

Diseases which afflict poultry such as chicken, turkey and goose are termed as poultry diseases. Some of the common poultry diseases are erysipelas, tick fever and tick paralysis, avian encephalomyelitis, virulent Newcastle disease, fowl pox, haemoproteus, Marek's Disease, etc. Poultry health management aims to prevent and control these diseases. This chapter discusses in detail these poultry diseases and the ways of managing them.

Edith Powell

1

Introduction

Poultry science is involved in the study of various types of birds which are raised for eggs, meat and feathers. It also studies the processes of incubation and hatching, as well as poultry products, poultry breeding and poultry processing. This chapter has been carefully written to provide an easy understanding of these facets of poultry science.

POULTRY SCIENCE

Poultry science is the study of all types of birds raised for food and feathers, including chickens, turkeys, geese and ducks. Poultry scientists work to improve poultry production and food safety. Poultry science is a specialization of animal science concerned with finding improved and more effective ways of raising poultry that produce flavorful and healthy eggs and meat. Poultry scientists may also seek to increase poultry production and profit while maintaining quality.

Poultry is any domesticated bird used for food. Varieties include chicken, turkey, goose, duck, Rock Cornish hens, and game birds such as pheasant, squab and guinea fowl. Also included are huge birds such as ostrich, emu and rhea (ratites). The term poultry actually refers to a variety of bird types raised on farms for food, fiber, or entertainment. Poultry is a big part of the diet. We consume it in many forms—meat, eggs, broth—and it is included in much of the food we eat. Poultry is big industry.

Broodiness

Broodiness is the action or behavioral tendency to sit on a clutch of eggs to incubate them, often requiring the non-expression of many other behaviors including feeding and drinking. Being broody has been defined as "Being in a state of readiness to brood eggs that is characterized by cessation of laying and by marked changes in behavior and physiology". Broody birds often pluck feathers from their chest and abdomen, using them to cover the eggs. As a consequence of this, they develop one or several patches of bare skin on the ventral surface. These reddish, well-vascularized areas of skin are usually called brood patches, and improve heat transfer to the eggs. Broodiness is usually associated with female birds, although males of some bird species become broody and some non-avian animals also show broodiness.

A brooding white tern (Gygis alba).

In Domestic Poultry

Broody hens can be recognized by their behavior. They sit firmly over the eggs, and when people approach or try to remove the eggs, threaten the person by erecting their feathers, emitting a characteristic sound like *clo-clo-clo* and will peck aggressively. When broody, hens often temporarily cease eating or reduce their feed consumption.

Letting eggs accumulate in a relatively dark place near the floor often stimulates hens to become broody. Placing artificial eggs into nests also stimulates broodiness. Keeping hens in dark places with warm temperatures and in view of vocalising orphan chicks can induce broodiness, even in breeds that normally do not go broody.

Some environmental conditions stimulate broodiness. In heavy breeds of chickens, warm weather tends to bring about broodiness. Removing eggs each day, out of the sight of the hens, helps avoid broodiness not only in domestic poultry but also in some wild species in captivity. This continued egg laying means more eggs are laid than would occur under natural conditions. Poultry farming in battery cages also helps to avoid broodiness.

A brooding domestic hen.

In Commercial Egg-laying

Because hens stop laying when they become broody, commercial poultry breeders perceive broodiness not as a normal physiological process, but as an impediment to egg and poultry meat production. With domestication, it has become more profitable to incubate eggs artificially, while keeping hens in full egg production. To help achieve this, there has been intense artificial selection for non-broodiness in commercial egg laying chickens and parent stock of poultry. As a result of this artificial selection, broodiness has been reduced to very low levels in present-day breeds of commercial fowl, both among egg-laying and meat-producing breeds.

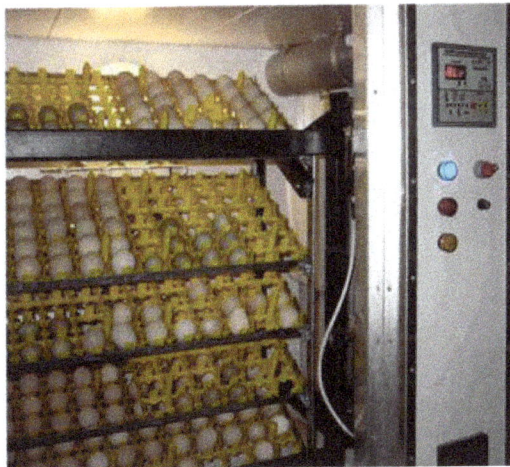

An egg incubator.

Physiological Basis

Broodiness is due to the secretion of the hormone prolactin by the anterior lobe of the hypophysis. Prolactin injection in hens provokes egg laying to stop within a few days, vitellum reabsorption, ovary regression (hens only have a left ovary) and finally broodiness. However, attempts to stop broodiness by the administration of several hormones have failed because this state, once evoked, requires time to revert.

Prolactin injections inhibit the production of gonadotropin hormone, a hormone that stimulates ovarian follicles which is produced in the frontal lobe of hypophysis. Castrated males can go broody with baby chicks, showing that broodiness is not limited to females, however, castrated males do not incubate eggs.

Contrary to common opinion, the temperature of broody hens barely differs from that of laying hens. Broody hens pluck feathers from their chest, using them to cover the eggs. As a consequence of this, they develop one or several patches of bare skin on the ventral surface. These reddish, well-vascularized areas of skin are usually called brood patches. which improve heat transfer to the eggs.

Genetic Basis

Broodiness is more common in some chicken breeds than others, indicating that it is a heritable characteristic. Breeds such as Cochin, Cornish and Silkie exhibit a tendency to broodiness, including brooding eggs from other species such as quails, pheasants, turkeys and geese. In some breeds such as the White Leghorn, broodiness is extremely rare.

Some studies on crosses of chicken breeds point to the hypothesis of complementary genes acting on broodiness. Other results point to the hypothesis of sex-linked genes, or, inheritance through the maternal chromosome. Although these studies have been made on different breeds of chickens, their results are not contradictory. There is common agreement that artificial selection for egg production succeeded in reducing the incidence of broody hens in chicken populations.

Chicken breeds that commonly exhibit broodiness:

- Cochin,
- Manx Rumpy,
- Plymouth Rock,
- Silkie,
- Sussex,
- Saipan Jungle Fowl,
- Orpington,
- Kraienköppe,
- Belgian Bearded d'Anvers,
- Icelandic Chicken,

- Java (chicken),
- Philippine Native Chicken,
- Belgian Bearded d'Uccle,
- Iowa Blue,
- Nankin,
- Delaware,
- Booted Bantam,
- New Hampshire,
- Pekin,
- Dutch Bantam,
- Indian Game (Cornish).

Incubation and Hatching

Incubation is the act Forced Draft Incubators:

- Based on heating source:
  - Hot air incubator,
  - Hot water incubator.
- Based on fuel used:
  - Gas operated incubator,
  - Oil operated incubator.

Location

The chick hatcheries are modern buildings that provide separate rooms for each hatchery operations, but each room has its individual requirements. The hatchery area should be a separate unit with its own entrance and exit, unassociated with those of the poultry farm. The hatchery should be situated at least 1000 ft from poultry houses to prevent horizontal transmission of disease-producing organisms from the chicken houses to the hatchery.

Size of the Hatchery

The size of the hatchery is based on the egg capacity of the setters and hatchers, number of eggs that can be set each week and number of chicks hatched each week. Also, necessary space to be allotted for future expansion.

Hatchery Design

Hatchery should be constructed in such a manner that the hatching eggs may be taken in one end and the chicks removed at the other. In other words, eggs and chicks should flow through the hatchery from one room to the one next needed in the hatching process. There should not be no backtracking. Such a flow affords better isolation of the rooms and there is less human traffic throughout the building.

Hatchery Construction

Hatchery buildings should be intricately designed, properly constructed, and adequately ventilated. Brief general points to be considered are:

- Width of the hatchery: The width of the setter and hatcher rooms is to be determined by the type of the incubator used. Find the depth of the incubators; then allow space for the working aisles, behind the machines and the walls.

- Height of the ceiling: The height of the ceiling should be at least 10 ft.

- Walls: Fireproof material should be used in constructing the walls of the hatchery also prevents the growth of molds common to walls that are porous and absorbent.

- Ceiling material: Most hatchery rooms have a high humidity, and during cold weather condensation of moisture on the ceilings is common. Hence, the ceiling material is to be waterproof.

- Doors: The hatchery doors are wide enough for easy movement of trolleys, chick boxes etc. The door openings should be 8 ft high and at least 4 ft wide, and doors double-swinging.

- Floor: All floors must be concrete, preferably with imbedded steel to prevent cracking. The concrete must be given a glazed finish. Slope of the floor should never be greater than 0.5 inch in 10 feet.

Hatchery Rooms/Structures

Hatchery rooms must be adequate in size. Usually, hatcheries of medium size will hatch chicks twice a week, but large hatcheries will hatch more than two hatches per week. Consequently, hatching schedules will affect the size of some rooms in the hatchery.

Shower Room

To maintain bio-security it is essential that all persons entering the premises shower and change into clean clothing in an adjoining room. It is the only entrance and exit, and the hatchery becomes an isolated unit as far as human beings are concerned.

Hatching Eggs Receiving Counter

Employees delivering hatching eggs to the hatchery must not enter the hatchery in the course of their duties. Eggs should be delivered to the hatchery through a specialized door.

Fumigation Room

The fumigation room should be as small as possible in order to reduce the amount of fumigant used. A fan should be used to circulate the air and exhaust the fumigant.

Egg Holding (Egg-cooler) Room

Egg holding room should be about 8 ft high, insulated, slowly ventilated, with complete air movement, cooled, and humidified. The room must be refrigerated to maintain a temperature of 65 °C. A forced-air type of refrigeration unit is required in order to keep a uniform temperature throughout the room.

Pre Incubation Warming Room

Here eggs are kept for the purpose of drying the 'sweat' over eggs. It can be achieved by providing sufficient number of ceiling fans in this room.

Setter Room

Setters (incubators) are kept in this room. The size of the setter room will depend on the make of the equipment used. The incubating equipment takes relatively little floor space. The exact room size involves the aisle and working area necessary to move the eggs and chicks in and out of the machines. A minimum space of 3 ft should be allotted between the sides of adjacent setters and from wall to sides or back of the setters. Similarly, minimum of 10 ft should be allotted in front of two setters when kept face-to-face arrangements.

Egg Candling (Dark) Room

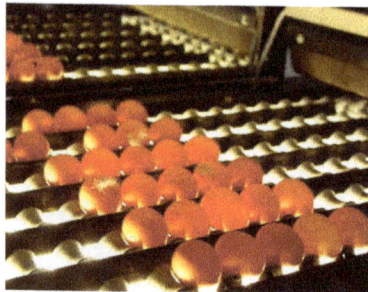

This room is usually constructed in between setter and hatcher room for candling eggs. Candling is usually practiced when eggs are transferred from setter to hatcher. Provisions should be made to dark the room to facilitate easy candling.

Hatcher Room

Hatchers are kept in this room. Here sufficient spaces are to be allowed around hatcher similar to that of setter room. Since it is prone for contamination with fluffs and debris at the time of hatching, the door towards setter room is to be tightly closed unless the necessity arises (at the time of egg transfer).

Chick Holding Room

Next to hatcher room, chick-holding room is present. A relative humidity of 65% is maintained to prevent excessive chick dehydration. Here, the chicks are sex-separated, graded, vaccinated and placed in chick boxes.

Wash Room

After chicks are boxed, the trays are washed in a tray washer in the washroom. Necessary pipelines with high-pressure pumps are kept in this room.

Clean Room

After the trays are washed, they are placed in their trolleys and moved to the adjacent clean room.

Principles of Incubation

Five major functions are involved in the incubation and hatching of poultry eggs. They are:

- Temperature,

- Humidity,

- Ventilation (Oxygen and Carbon dioxide level and air velocity),

- Position of eggs,

- Turning of eggs.

Temperature

Temperature is the most critical environmental concern during incubation because the developing embryo can only withstand small fluctuations during the period. Embryo starts developing when the temperature exceeds the Physiological Zero. Physiological zero is the temperature below which embryonic growth is arrested and above which it is reinitiated. The physiological zero for chicken eggs is about 75 °F (24 °C). The optimum temperature for chicken egg in the setter (for first 18 days) ranges from 99.50 to 99.75 °F and in the hatcher (last 3 days) is 98.5 °F.

Humidity

Incubation humidity determines the rate of moisture loss from eggs during incubation. In general, the humidity is recorded as relative humidity by comparing the temperatures recorded by wet-bulb and dry-bulb thermometers.

Recommended incubation relative humidity for the first 18 days ranging between 55 and 60% (in setter) and for the last 3 days ranging between 65 and 75%. Higher humidity during hatching period is given to avoid dehydration of chicks.

Ventilation

Ventilation is important in incubators and hatchers because fresh oxygenated air is needed for the respiration (oxygen intake and carbon dioxide given off) of developing embryos from egg setting until chick removal from the incubator.

The oxygen needs are small during the first few days compared to the latter stages of development. Oxygen content of the air at sea level is about 21%. Generally the oxygen content of the air in the setter remains at about 21%. For every 1% drop in oxygen there is 5% reduction in hatchability.

Carbon dioxide is a natural by-product of metabolic processes during embryonic development and is released through the shell. The tolerance level of CO_2 for the first 4 days in the setter is 0.3%. CO_2 levels above 0.5% in the setter reduce hatchability and completely lethal at 5.0%.

Since the normal oxygen and CO_2 concentrations present in air seem to represent an optimum gaseous environment for incubating eggs, no special provision to control these gases is necessary other than to maintain adequate circulation of fresh air at the proper temperature and humidity.

Position of Eggs

Artificially incubating eggs should be held with their large ends up. It is natural for the head of the chick to develop in the large end of the egg near the air cell, and for the developing embryo to orient itself so that the head is uppermost. When the eggs are incubated with the small end up, about 60% of the embryos will develop with the head near the small end. Thus, when the chick is ready to hatch, its beak cannot break into the air cell to initiate pulmonary respiration. Eggs positioned horizontally will incubate and hatch normally as long as they are turned frequently. Under normal circumstances eggs are set with large end up for the first 18 days (in setter) and in horizontal position for the last 3 days (in hatcher).

Turning of Eggs

Birds, including chickens and quail, turn their eggs during nest incubation. Nature provides nesting birds with the instinct of turning eggs during incubation. Similarly eggs to be turned at least 8 times a day. Turning of eggs during incubation prevents the developing embryo adhering to the extra-embryonic membranes and reduces the possibility of embryo mortality. In large commercial incubators the eggs are turned automatically each hour i.e. 24 times a day. Most eggs are turned to a position of 45° from vertical, and then reversed in the opposite direction to 45° from vertical. Rotation less than 45° are not adequate to achieve high hatchability. Turning is not required in Hatcher.

Factors	Setter	Hatcher
Temperature	99.50 to 99.75 °F	98.5 °F
Relative Humidity	55-60 %	65-70 %
Position	Large end up	Horizontal
Turning	Manual - 8 times Automatic - 24 times	No turning

Handling of Hatching Eggs and Storage

The quality of hatching egg cannot be improved after lay but one can reduce the loss in hatching egg quality by adopting some standard procedures.

Maintaining Egg Quality in the Breeder House

The hen will lay her eggs over nesting material. Use of enough clean, dry and mold-free nesting material can avoid cracked and dirty eggs. Similarly hens to be trained to use nests to lay eggs instead of laying on floors by providing sufficient number of nest boxes well in advance before the laying starts.

The frequency of hatching egg collection is very important to maintain quality. Hatching eggs should be collected at least 4 times a day. Hatching eggs are susceptible to contamination and every effort must be made to reduce the microbial load. Therefore, it is imperative that people wash and sanitize their hands before collecting eggs from the nests or egg belts. The flats that eggs are placed on must also be sanitized and free of organic material.

Selection of Hatching Eggs

Not all eggs laid by a breeding flock are set. Eggs that are cracked, dirty or misshapen are usually not used for hatching. Very small or very large eggs do not hatch as well as eggs in the middle size range. Eggs with thin or very porous shells are not likely to hatch well because of excessive losses of water during incubation.

Reducing Contamination of Hatching Eggs

Poor hatching egg sanitation can be a major cause of lower hatchability and poor chick quality. There is no such thing as a sterile eggshell. More bacteria are picked up on the shell when the egg passes through the cloaca where urine and intestinal contents also pass. The bacterial load found on an eggshell at the time of lay ranges from 300 to 500 organisms. After oviposition, every surface the egg comes in contact with can further inoculate the shell surface. After an egg is laid it begins to cool. During the cooling process the egg contents begin to shrink and producing negative pressure. This is one of the more opportune times for bacteria on the shell surface to penetrate the eggshell.

Egg has many natural defense mechanisms to reduce bacterial penetration. The shell itself provides some protection. The cuticle on the surface of the eggshell is the best natural barrier to penetration. The inner and outer shell membranes provide additional barriers. The albumen provides a somewhat effective control over contamination. The albumen has a high pH in which most bacteria cannot survive. The chalazae contain an enzyme, lysozyme, which has antibacterial properties.

Many breeder people choose some methods to reduce the microbial load over the eggshell. Sanding, buffing, and wiping the hatching eggs are not good methods of sanitation. Sanding and buffing will remove at least part of the cuticle resulting in eggs that are more susceptible to penetration. Fumigation with formaldehyde gas is an effective method of sanitizing hatching eggs. Solutions containing quaternary ammonium compounds, formalin, hydrogen peroxide or phenols may be moderately effective in reducing the microbial load over hatching eggs. DO NOT wash eggs unless necessary. If it is necessary to wash eggs always use a damp cloth with water warmer than the egg. This causes the egg to sweat the dirt out of the pores. Never use water cooler than the egg. Also, do not soak the eggs in water.

Storage of Hatching Eggs

In normal hatchery operations, eggs cannot be set immediately after they are laid. Many hatcheries set eggs once or twice in a week. If hatching eggs are stored up to 1 week, hatching eggs should be kept in an egg holding room with the temperature of 650 °F and the relative humidity of 75%. When storing eggs less than 10 days, store them with the large end up. If eggs are held for 10 days or more, hatchability will be improved if stored with small end up.

Hatchery Operations

The operation of a chick hatchery involves the production of the largest number of quality chicks possible from the hatching eggs received at the hatchery. In addition, chicks must be produced economically. The sequences of hatchery operations followed in commercial hatcheries are:

Advantages

- Securing hatching eggs,
- Traying hatching eggs,
- Fumigation,
- Cold Storage,

- Warm eggs prior to setting,
- Loading of eggs,
- Candling,
- Transfer of eggs,
- Pulling the hatch,
- Hardening,
- Grading,
- Sexing,
- Vaccination,
- Chick delivery,
- Washing and cleaning,
- Disposal of waste.

Securing Hatching Eggs

Hatcheries can get the hatching eggs from any one of the following ways:

- From own breeder flock,
- From other breeder flocks,
- From other hatcheries.

Traying Hatch Eggs

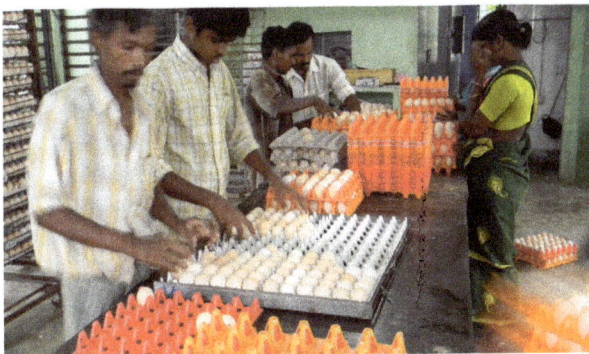

Traying hatch eggs.

The eggs from the breeder flocks should be transferred to the egg setter trays in the hatchery immediately after receiving.

Fumigation of Hatching Eggs

After traying, eggs are to be kept in the fumigation chamber for fumigation. Fumigating with 3x concentration of formaldehyde for 20 minutes will kill about 97.5 to 99.5% of the organisms on the shells. One 'x' concentration means 20 g of $KMnO_4$ with 40 ml of formalin for 100 cubic feet (3x means 60 g of $KmnO_4$ + 120 ml of formalin for 100 cubic feet).

Cold Storage

When the eggs are not set immediately after receiving, they should be kept in cold rook at the temperature of 65 °F and 75% relative humidity.

Warm Eggs Prior to Setting

Approximately 6 hours prior to placing eggs in the setter they should be moved from the egg-cooler room to normal room temperature. Here, atmospheric air condenses over eggshell and form water droplets over eggshell, which is called as 'Sweating'. It is advantageous to warm eggs before placing them in the incubator by avoiding creation of low temperature in the machine by placing cool eggs directly.

Loading of Eggs

Placing of eggs in the setter is called 'Loading of eggs'. Eggs can be set in the setter either all-in all-out basis or batch basis. Most of the commercial hatcheries are practicing batch system of loading eggs in the setter that will minimize the initial time taken to reach normal incubation temperature in the setter. In this case, each setter is having hatching eggs with different stages of embryonic developments.

Candling

Candling is a process in which eggs are kept in front of a light source to find out the defects in eggshell, embryonic development etc. Candling can be done as early as five days of incubation, but errors in candling often occur at this time. Under commercial operations, candling is done when the eggs are transferred from setter to hatcher (at 19th day for chicken eggs). There are two methods of candling that can be used. The fastest method involves the use of a table or mass candler. An entire tray of hatching eggs may be placed on the mass candler and examined with one observation. Candling with a spot candler or individual candler is a little slower, but it is more accurate.

Transfer of Eggs

In modern incubators, eggs are transferred from setter to hatcher at 19th day of incubation (for chicken egg) or when approximately 1% of the eggs are slightly pipped. In general, one-seventh of total incubation period is needed to keep eggs in the hatcher.

Pulling the Hatch

The process of removing the chicks from the hatcher is often called pulling the hatch. Chicks should be removed from the hatcher as soon as all are hatched and about 95% are dry. Excessive drying in the hatcher should be avoided.

Hardening the Chicks

When the chicks are first placed in the chick boxes they are soft in the abdomen, are not completely fluffed out, and do not stand well. They must be "hardened" by leaving them in the boxes for 4 or 5 hours. Such hardening makes it easier to grade the chicks for quality, and the chicks are more easily vent-sexed.

Grading the Chicks

No chick below the minimum standard must be allowed to go to a customer. Some standards for quality are, 1) No chick deformities 2) No unhealed navels 3) Above a minimum weight 4) Not dehydrated and 5) Stand up well.

Sexing the Chicks

Sexing the chicks.

Layer type day-old chicks are need to be sex separated either by vent sexing or auto-sexing (feather sexing). In case of meat-type day-old chicks sexing is not practiced.

Vaccination

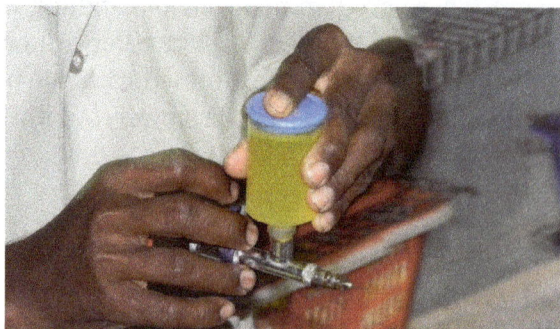
Vaccination

Most chicks are vaccinated against Marek's disease in hatchery before delivery. Most common method of vaccination of day-old chicks is by subcutaneous method in the nape of the neck.

Chick Delivery

Baby chicks should reach the customer's farm early in the morning. Not only the weather is cooler during this part of the day but the early arrival allows a full day for close observation of the chicks by the caretaker.

Washing and Cleaning

Washing and cleaning.

Cleaning the hatchery between hatches is of primary importance. The process must be complete. Except for the setters and setter room, every piece of equipment must be thoroughly vacuumed, scrubbed, disinfected and fumigated.

Disposal of waste

Hatchery wastes include infertile and non-hatched eggs, and dead and cull chicks that should be disposed in such a manner not to create problem to the neighbors and also not to contaminate the hatchery premises.

Disposal of waste.

Steps Involved in Commercial Hatchery Operations

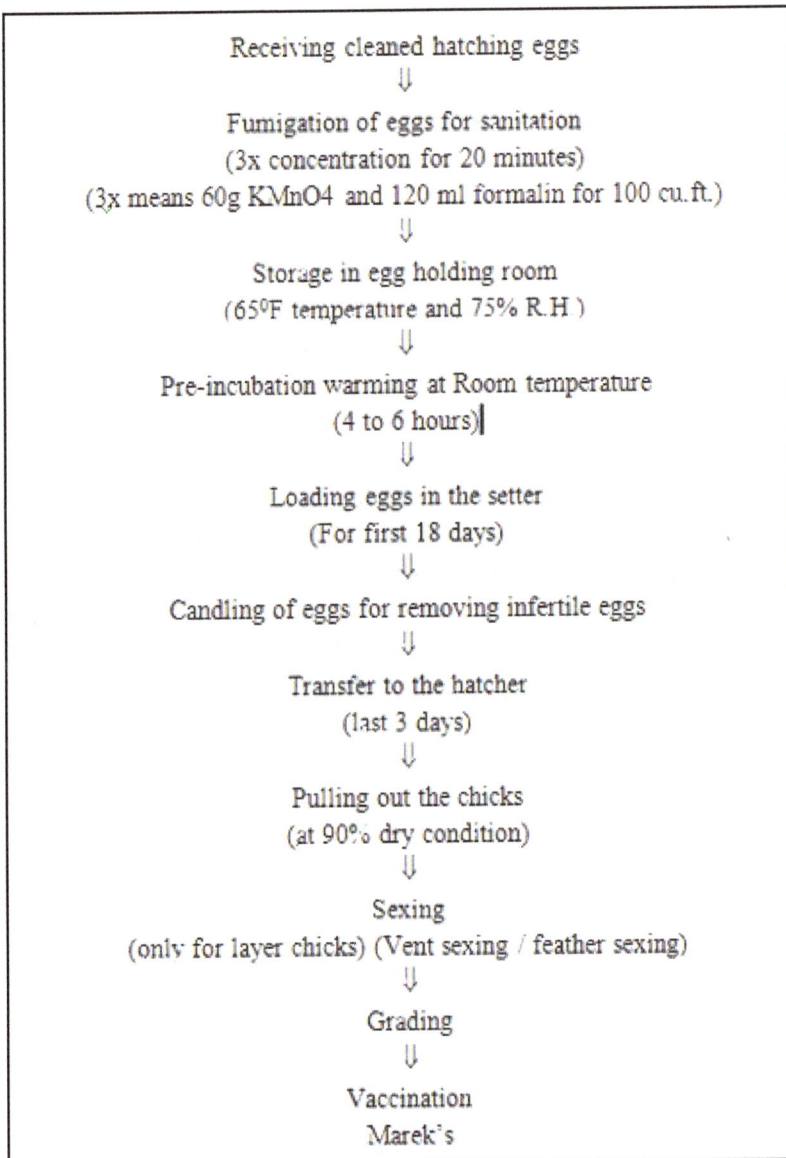

Receiving cleaned hatching eggs
⇓
Fumigation of eggs for sanitation
(3x concentration for 20 minutes)
(3x means 60g KMnO4 and 120 ml formalin for 100 cu.ft.)
⇓
Storage in egg holding room
(65°F temperature and 75% R.H)
⇓
Pre-incubation warming at Room temperature
(4 to 6 hours)
⇓
Loading eggs in the setter
(For first 18 days)
⇓
Candling of eggs for removing infertile eggs
⇓
Transfer to the hatcher
(last 3 days)
⇓
Pulling out the chicks
(at 90% dry condition)
⇓
Sexing
(only for layer chicks) (Vent sexing / feather sexing)
⇓
Grading
⇓
Vaccination
Marek's

POULTRY LITTER

The poultry industry, one of the largest and fastest growing agro-based industries, is not only providing a huge amount of animal protein but also producing millions of tons of waste materials whose proper utilization can bring a great economic and environmental advantage. Having numerous beneficial potentialities, in many cases poultry litter and manure are being used inappropriately and causing many economic and environmental hazards.

Poultry manure or chicken manure is the organic waste from poultry composed of mainly feces and urine of chickens. The mixture of poultry manure with spilled feed, feathers and bedding materials like wood shavings or sawdust is referred as poultry litter. Poultry litter is an organic manure enriched with many major plant nutrients like N, P, K and many trace elements like Zn, Cu, As etc. The composition and quality of a poultry litter varies with the types of poultry, types of litter used, diet and dietary supplements, and collection and storage of the litter.

Use of Poultry Litter

As a Good Fertilizer

Poultry litter is a rich fertilizer with a great number of nutrients essential for plant growth and has been used as organic fertilizer for centuries. It contains a high amount of major nutrients especially N, P, and K. Trace nutrients like Cu, Zn, As etc. are also present in this manure. Bird feces largely add all these nutrients to poultry manure. The manure release nutrients to the soil for plant uptake upon their decomposition in the soil. Poultry litter manure is found to give very good crop productions in various experiments. For instance, corn yields were obtained higher in a number of studies due to the application of poultry manure. Fertilizer studies with cotton in the U.S. showed that poultry litter is a valuable source of plant nutrients supplying N and metals Fe, Cu, Zn and Mn. It is locally readily available fertilizer at a low cost and can be very useful in reducing vegetable production cost.

As an Effective Soil Amendment

Excessive application of chemical fertilizers and continuous cultivation of the same crop in the particular field cause deterioration in soil structure and overall soil quality. Different experiments showed that addition of poultry manure and litter improves many vital properties of soils. Poultry manure has been found to decrease the bulk density and to increase the water holding capacity, organic matter content, oxygen diffusion rate, and aggregate stability of the soils.

When poultry litter is used as a mulching material, it conserves soil moisture and save the surface feeding roots from drying out in the summer heat. Poultry manure is capable of improving the biological fertility of mine tailings. Poultry litter is increasingly being used in the rehabilitation of disturbed land resulting from mining and other industrial activities.

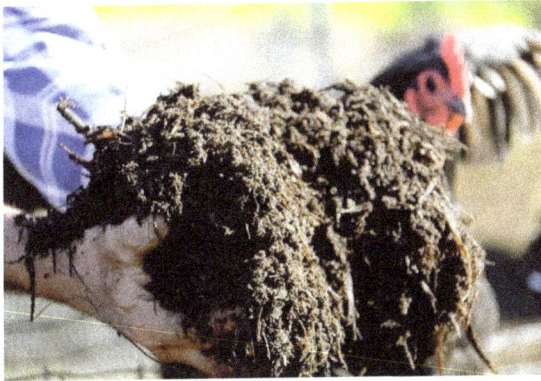

As a Good Animal Feed

Poultry manure or litter can be used as feed for cattle and fish. To use the litter as feed the foreign materials in it like plastic, feather, glass etc. must be removed. Poultry litter with ash content more than 28% is not safe to be used as feed, so low ash content must be maintained. Unprocessed poultry litter contains pathogenic microorganisms like Clostridium, Salmonella and Enterobacter spp.; and feed additives, which are added in poultry diet and can be present in the litter as waste by-products, such as antibiotics, coccidiostats, and arsenicals. Therefore to use as feed, proper processing is a must to eliminate these substances from the litter. Properly processed poultry manure and litter are enriched with protein, minerals and fiber and have been mixed with cattle feed for these nutrients. In U.S., they have been used as a useful feed ingredient for around 40 years.

As a Good Fuel Source

Poultry litter can be used as a great source of fuel. Poultry litter with moisture content less than 15% can be burnt directly as fuel to generate heat energy. Biogas, a very useful combustible gas with around 60% methane, can be produced by anaerobic digestion of poultry waste. Biogas, produced by poultry litter, can be used for various purposes including fuel for engines, to produce electricity, to produce heat by simply burning and so on. Many states in U.S. are producing a huge amount of turkey and broiler waste to use those as renewable green sources to produce electricity.

Other Uses

Using properly processed poultry litter for various purposes brings some other advantages-

- Reduces the quantity of useless wastes which needs to be disposed.

- Reduces odor problems.

- Proper using of litter protects the groundwater from potential litter contamination.

- Create an income source for the people involved in litter processing.

- Kills harmful microbes, flies and weed seeds present in the litter.

- Produce energy from waste.

How to use Poultry Litter

Composting

Using fresh poultry litter is not safe as it contains many pathogenic organisms, harmful chemicals, and weed seeds; therefore, composting is necessary to use them as fertilizer. To compost, fresh chicken manure should be mixed with different ratios of Carbon source. The Carbon source is also needed to be a bulking agent that facilitates the aeration of the compost pile. The pile temperature should be maintained between 55 and 65 °C for minimum 2 weeks.

Turning the pile regularly is needed to homogenize the pile. Turning and maintaining pasteurizing heat (55-65 °C) for 2 weeks kills all weeds and pathogens and gives compost which is not properly mature. The compost thus results is not stable and emits strong odor but still that is good fertilizer with no weeds and pathogens.

Some special cares are needed to produce good quality litter compost:

- To avoid leaching and potential groundwater contamination, composting should be done on an impermeable base.

- An elevated area or a building should be used for composting to deter extraneous runoff entering the compost pile.

- To prevent nutrient leaching and contamination, the compost pile is needed to be secured from rain.

- To avoid emitting odorous gasses, the pile should be protected from the wind.

- Maintaining optimum moisture content is useful to avoid dust and odor emission. When the pile is too dry, excessive dust may be produced and when the pile is too wet, an excessive odor is emitted. Maintaining moisture content around 50-55% is the best way avoiding these.

Land Application

Some strategies must be followed to apply poultry litter as a fertilizer:

- Poultry compost works best during the active growth of the plants or immediately before planting, that's that should be added at that time. Application of poultry litter during fall or winter is not a good choice, as crops cannot properly utilize the nutrients at that time.

- Timing of application of poultry litter should be determined on the basis of maximizing crop recovery of nutrients.

- Poultry litter application should be done on the basis of balanced nutrient requirements of the crops.

- To avoid gaseous losses of N and runoff losses of other nutrients, poultry litter should be incorporated into the soil.

Environmental Monitoring

To ensure sustainable production and healthy environment, environmental monitoring is a vital component of a balanced poultry litter management practice. Regular analysis of manure samples, soil and drainage water sampling and their subsequent analysis for contaminants and nutrients content, maintaining records of all farm activities are the important parts of environmental monitoring for poultry manure application.

POULTRY PRODUCTION AND PRODUCTS

Poultry meat and eggs are among the animal-source foods most widely eaten at global level, across greatly diverse cultures, traditions and religions. Consumption of poultry meat and eggs – and of animal-source foods in general – has increased rapidly in past decades. Growing demand has mostly been driven by population growth, urbanization and rising incomes in developing countries. Chicken dominates meat consumption as it is generally affordable, low in fat and faces few religious and cultural barriers.

Demand for poultry meat and eggs is expected to continue increasing due to population growth and rising individual consumption. The market for poultry meat is projected to increase regardless of region or income level, with per capita growth slightly higher in developing than in developed regions.

Poultry meat and eggs contribute to human nutrition by providing high-quality protein and low levels of fat, with a desirable fatty acid profile. Urban and peri-urban dwellers generally eat poultry raised in intensive systems, either locally produced or imported, but niche markets exist for indigenous poultry and poultry products. In rural areas of developing counties, most households consume meat and eggs from their own, usually small, flocks of indigenous birds.

Meat and eggs are not the only important poultry products. A significant by-product is manure, which has robust economic value, whether sold or directly applied to crops by farmers. Down and feathers can also be sold. In mixed farming systems, other products such as egg shells can be fed to other farm animals (e.g. pigs).

Poultry in Human Nutrition

Besides being rich in protein, poultry meat is good source of phosphorus and other minerals, and of B-complex vitamins. It contains less fat than most cuts of beef and pork. Poultry meat is low in harmful trans fats, but high in beneficial monounsaturated fats – which make up about half of the total. Eggs are a good source of high biological-value protein and easily digestible.

Because of their inadequate diets, poor people living in low-income regions such as in sub-Saharan Africa and South Asia are particularly vulnerable to a number of diseases. Eating more poultry meat and eggs can substantially benefit such people, especially pregnant women, children and the elderly. There is a growing evidence that poultry meat can make a significant difference in fighting child malnutrition.

The question of cholesterol in eggs – which prompted a decline in consumption in many developed countries – was once seen as an important issue but has now been largely superseded. Attempts to reduce cholesterol in eggs have not only proved impractical but are considered unnecessary these days. Consumption of one or two hen's eggs a day is no longer considered a risk factor to human health for most of the population.

Collection and Transport

Efficient transport of poultry and poultry products is essential as poor transportation can lead to significant loss of quality and production and/or seriously harm the health and welfare of poultry.

Large-scale or medium-sized poultry producers are generally located close to processing plants or markets and contract or own means of transport for live birds. Birds that are slaughtered in abattoirs are often trucked live in crates.

Processed poultry carcasses, parts or products (including eggs) are usually shipped in refrigerated trucks. Recent developments in transportation, such as long-distance cold-chain shipments, have made it possible to trade and transport poultry and poultry products over long distances.

In many developing countries, however, bad roads and lack of marketing, refrigeration, and other infrastructure still hold back the spread of commercial poultry production. Poultry and eggs are often transported on bicycles, motorcycles, cars, horses, or even on foot. Eggs may be packaged in banana leaves or sawdust, and birds carried in baskets. Given such rudimentary transport conditions, poultry may reach market in bad shape, often with broken feathers and bruises.

Processing Systems

Large-scale processors operate their own abattoirs where poultry are slaughtered, processed, graded, packaged, stored and distributed either for direct sale or under contract to other large organizations such as supermarkets. While whole birds are the leading items sold, poultry meat can be further processed into various products from simple cuts to oven-ready meals. In recent decades, consumer preference has shifted from fresh, whole birds to cut-up bird parts and convenience products.

In most small-scale abattoirs in developing countries, slaughter is carried out manually using simple tools. After being killed, birds are scalded in hot water and then plucked and eviscerated, mostly by hand. The meat may then be sold dressed (slaughtered, bled and plucked); eviscerated and ready to cook; in poultry parts (legs, wings, etc.); or deboned (muscle, fat and skin only).

Eggs may be sold either as table eggs or processed into egg products that go into a wide range of food products, including soups, sauces, cakes, biscuits and desserts. Making egg products involves breaking the shells, filtering, mixing, stabilizing, blending, pasteurizing, cooling, freezing or drying, and finally packaging.

Risks to Human Health

Poultry production is one of the fastest-growing livestock subsectors, and this raises a number of health issues, not only for the people working with poultry but also for those consuming poultry products.

Most humans, in fact, often come into contact with poultry, either directly or indirectly.

The greatest concern arising from such proximity is the risk of infection with the highly pathogenic avian influenza (HPAI) virus. Handling and slaughtering live, infected poultry poses the greatest hazard, but transmission can also occur through contact with droppings, feathers, organs and blood.

The consumption and handling of contaminated poultry meat and eggs can lead to food poisoning in people. The main causes of human intestinal infections from this source are bacteria, principally Salmonella and Campylobacter. Another threat to human health lies in the inappropriate use of antimicrobials in poultry production, leading to the development of antimicrobial-resistant microorganisms.

Effective control systems and policies are critical in ensuring product safety and in reducing risks to human health.

POULTRY PROCESSING

Poultry processing is the preparation of meat from various types of fowl for consumption by humans.

Poultry is a major source of consumable animal protein. For example, per capita consumption of poultry in the United States has more than quadrupled since the end of World War II, as the industry developed a highly efficient production system. Chickens and turkeys are the most common sources of poultry; however, other commercially available poultry meats come from ducks, geese, pigeons, quails, pheasants, ostriches, and emus.

Characteristics of Poultry

Poultry is derived from the skeletal muscles of various birds and is a good source of protein, fat, and vitamins and minerals in the diet.

Classification of Birds

Birds bred for poultry production are generally grown for a particular amount of time or until they reach a specific weight. Rock Cornish hens, narrowly defined, are a hybrid cross specifically bred to produce small roasters; in the marketplace, however, the term is used to denote a small bird, five to six weeks old, that is often served whole and stuffed. Seven-week-old chickens are classified as broilers or fryers, and those that are 14 weeks old as roasters.

Fat Content

The fat content of poultry differs in several ways from that found in red meat. Poultry has a higher proportion of unsaturated fatty acids compared with saturated fatty acids.

Both turkey and chicken contain about 30 percent saturated, 43 percent monounsaturated, and 22 percent polyunsaturated fatty acids. The high levels of unsaturated fatty acids make poultry more susceptible to rancidity through the oxidation of the double bonds in the unsaturated fatty acids. Saturated fatty acids, on the other hand, do not contain double bonds in their hydrocarbon chains and are resistant to oxidation. However, this fatty acid ratio has led to the suggestion that poultry may be a more healthful alternative to red meat.

In birds fat is primarily deposited under the skin or in the abdominal cavity. Therefore, a significant amount of the fat can be removed from poultry by removing the skin before eating.

Microbial Contamination

Poultry provides an excellent medium for the growth of microorganisms. The principal spoilage bacteria found on poultry include Pseudomonas, Staphylococcus, Micrococcus, Acinetobacter, and Moraxella. In addition, poultry often supports the growth of certain pathogenic (disease-causing) bacteria, such as Salmonella.

Potential causes of contamination of poultry during the slaughtering and processing procedures include contact of the carcass with body parts that contain a high microbial load (e.g., feathers, feet, intestinal contents), use of contaminated equipment, and physical manipulation of the meat (e.g., deboning, grinding). Prevention of microbial contamination involves careful regulation and monitoring of the slaughtering and processing plants, proper handling and storage, and adequate cooking of raw and processed poultry products.

Slaughtering Procedures

Preslaughter Handling

When the birds have reached "harvest" time, they are generally taken off of feed and water. This allows their digestive tracts to empty and reduces the potential for contamination during processing.

At night the birds are caught by specially trained crews and placed into plastic or wooden transport cages. The birds are then transported to the slaughterhouse, where the trucks are often kept between sets of fans to ventilate the cages. In the next step the birds are removed from the cages and transferred to continuously moving shackles where they are suspended by both legs. The transfer is often done in a dark room illuminated by a red light; the birds are not sensitive to the red light and this helps to keep them calm.

The handling and transfer of birds both on the farm and at the slaughterhouse can be stressful. Stress can have negative effects on the quality of the final meat product, and therefore efforts are constantly being made to improve the preslaughter processes.

Slaughtering

Stunning and Killing

After the birds have been transferred to the moving shackles, they are usually stunned by running their heads through a water bath that conducts an electric current. Stunning produces unconsciousness, but it does not kill the birds. The birds are killed either by hand or by a mechanical rotary knife that cuts the jugular veins and the carotid arteries at the neck. Any birds not killed by the machine are quickly killed by a person with a knife assigned to the bleed area. The birds are permitted to bleed for a fixed amount of time, depending on size and species (e.g., 1 1/2 minutes for broilers). Any bird that is not properly bled will be noticeably redder after feather removal and will be condemned.

Scalding

Following bleeding, the birds go through scalding tanks. These tanks contain hot water that softens the skin so that the feathers can be removed. The temperature of the water is carefully controlled. If retention of the yellow skin colour is desired, a soft-scald is used (about 50 °C, or 122 °F). If a white bird is desired, a higher scald temperature is used, resulting in the removal of the yellow pellicle. Turkeys and spent hens (egg-laying birds that have finished their laying cycles) are generally run at higher temperatures—59 to 60 °C (138 to 140 °F).

Defeathering

The carcasses then go through the feather-picking machines, which are equipped with rubber "fingers" specifically designed to beat off the feathers. The carcasses are moved through a sequence of machines, each optimized for removing different sets of feathers. At this point the carcasses are usually singed by passing through a flame that burns off any remaining feathers.

An extra process, called wax dipping, is often used for waterfowl, since their feathers are more difficult to remove. Following the mechanical feather picking, the carcasses are dipped in a melted, dark-coloured wax. The wax is allowed to harden and then is peeled away, pulling out the feathers at the same time. The wax is reheated and the feathers are filtered out so that the wax can be reused. This process is usually performed twice.

The blood and feathers accumulated during these early steps are generally collected and rendered to make blood meal and feather meal. The feathers from ducks and geese are often carefully collected and used for down production.

Removal of Heads and Legs

The heads of the birds go into a channel where they are pulled off mechanically; the legs of the birds are removed with a rotary knife (much like a meat slicer) either at the hock or slightly below it, depending on national custom. The carcasses drop off the shackle and are rehung by their hock onto the eviscerating shackle line. By law in the United States, the scalding and defeathering steps must be separated by a wall from the evisceration steps in order to minimize cross-contamination.

Evisceration and Inspection

At this point the preen, or oil, gland is removed from the tail and the vent is opened so that the viscera (internal organs) can be removed. Evisceration can be done either by hand (with knives) or by using complex, fully automated mechanical devices. Automated evisceration lines can operate at a rate of about 70 birds per minute. The equipment is cleaned (with relatively high levels of chlorine) after each bird.

The carcasses are generally inspected during the evisceration process. The inspection procedures in the poultry industry vary around the world and may be performed by government inspectors, veterinarians, or plant personnel, depending on a country's laws. For example, in the United States the viscera are removed and placed on the side of the bird. Inspectors from the U.S. Department of Agriculture then examine the entire bird. The plant provides each inspector with an assistant who carries out any adjustments required by the inspector (e.g., removing the entire bird or removing some part of the bird). The rejected parts are placed in a container marked "inedibles," and the contents are generally dyed (often a blue-purple), under supervision of the inspector, in order to prevent possible mixing with edible parts.

Following inspection, the carcasses are further cleaned. The viscera are separated from the carcasses, and the edible offal are removed from the inedible offal. The heart, stomach, and liver are all considered edible offal and are independently processed. Stomachs are generally cut open and the inside yellow lining of the stomach along with the stomach contents are removed.

The lungs and kidneys are removed separately from the other visceral organs using a vacuum pipe. A final inspection is often carried out at this point, and the carcasses are then washed thoroughly.

Chilling

After the carcasses have been washed, they are chilled to a temperature below 4 °C (40 °F). The two main methods for chilling poultry are water chilling and air chilling.

Water Chilling

Water chilling is used throughout North America and involves a prechilling step in which a countercurrent flow of cold water is used to lower the temperature of the carcasses. The carcasses are then moved into a chiller—a large tank specifically designed to move the carcasses through in a specific amount of time. Two tanks are used to minimize cross-contamination.

A specified overflow of water for each tank is required by law in the United States and Canada. Although this renders the chilling process very water-intensive, it helps to minimize bacterial cross-contamination by diluting the microorganisms washed off the carcasses, thereby preventing recontamination.

Water chilling leads to an increase in poultry weight, and the amount of water gained is carefully regulated. In the United States the legal limits for water pickup are 8 percent for birds going directly to market and 12 percent for birds that will be further processed (the assumption is that they will lose the extra 4 percent by the time they reach the consumer).

Air Chilling

Air chilling is the standard in Europe. The carcasses are hung by shackles and moved through coolers with rapidly moving air. The process is less energy-efficient than water chilling, and the birds lose weight because of dehydration. Air chilling prevents cross-contamination between birds. However, if a single bird contains a high number of pathogens, this pathogen count will remain on the bird. Thus, water chilling may actually result in a lower overall bacterial load, because many of the pathogens are discarded in the water.

The final temperature of the carcasses before shipment is usually about −2 to −1 °C (28 to 30 °F), just above the freezing point for poultry. In some cases a slight crusting on the surface occurs during the final chilling. For water-chilled carcasses this final chilling takes place after packaging, when the carcasses are placed in an air chiller.

Processing of Poultry

Raw Poultry products

Whole or individual parts of birds may be packaged raw for direct sale. Poultry packaged in the United States must include instructions about safe handling, including the need to wash all equipment that has come in contact with raw poultry and the need to wash one's hands before preparing other foods. Most raw turkey is sold frozen, while most chicken is sold fresh.

Fresh Poultry

The birds are generally cut into a number of pieces, which are placed on plastic foam trays and covered with a plastic film. A "diaper" (absorbent paper with a plastic backing) is often used to catch any liquid that may be released from the meat. Fresh poultry should be used within 14 to 21 days after slaughter and generally should not be kept in the home refrigerator for more than three days. In the United States, poultry that has been frozen to a temperature of −5 to −4 °C (22 to 24 °F) and then allowed to thaw can legally be sold as "fresh."

Frozen Poultry

Most frozen poultry is vacuum-packed in plastic bags and then frozen in high-velocity freezers. The birds are kept in cold storage until needed. Before freezing, poultry may be injected with various salts, flavourings, and oils in order to increase the juiciness of the meat. Injections are usually done with a multi-needle automatic injector, and information about the added ingredients is indicated on the package label.

Frozen storage time (including poultry bought fresh and frozen in a home freezer) depends on the temperature of the freezer, the quality of the packaging, and the cycling of the freezer. For best results poultry should be used within three months. Frozen poultry products can be used directly in the frozen state or thawed first. Thawing should be done in the refrigerator or under running cold water to minimize the potential for microbial contamination.

Processed Poultry Products

Poultry may be further processed into other products. The number of processed poultry products has increased dramatically since the 1970s because of the low cost of poultry and its versatile, bland flavour.

Battering and Breading

Some poultry products are battered (e.g., with beer batter) or battered and breaded (e.g., with cracker meal, bread crumbs, or cornmeal) for frying. The meat may be either cooked or raw prior to coating. For battered and breaded poultry, the pieces are passed through a flour-based batter containing leavening and then through the breading ingredients. Many types of baked breadings have been developed to meet different tastes (e.g., Cajun or Japanese). To hold the breading to the poultry, the product is deep-fried for a short time. If the poultry is fully cooked in this process, the consumer will only have to heat the product before eating it. Chicken nuggets are a battered and breaded product that is marinated before coating.

Tumbling and Massaging

In the manufacturing of many poultry products, the meat is mixed with a variety of nonmeat ingredients, including flavourings, spices, and salt. Tumbling and massaging are gentle methods that produce a uniform meat mixture. A tumbler is a slowly rotating drum that works the meat into a smooth mixture. A massager is a large mixing chamber that contains a number of internal paddles. Cured turkey products (i.e., treated with sodium nitrite), such as turkey ham and turkey pastrami, are often tumbled or massaged during processing.

Smoking

Poultry may be smoked. Prior to smoking, the birds must be brined (soaked in a salt solution containing certain flavourings) and then allowed to dry. Smoking can be done using real wood shavings or a smoke flavouring. In the last case this must be labeled in the United States as "natural smoke flavor added."

Deboning and Grinding

Further processed poultry products leave the backs, necks, and bones available for their own processing. These materials are run through a machine called a mechanical deboner or a meat-bone separator. In general, the crushed meat and bones are continuously pressed against a screen and the edible, soft materials pushed through the screen. The resulting minced product is similar in texture to ground beef and has been used for many poultry products such as frankfurters (hot dogs) and bologna. Poultry frankfurters and bologna are made using a process similar to that for beef and pork. The meat is combined with water or ice, salt, and seasonings and chopped to emulsify the materials. The mixture is stuffed into plastic casings and cooked in a smokehouse. The meat is then quickly chilled, peeled, and vacuum-packaged. Bologna is stuffed into a larger casing and is not necessarily peeled.

POULTRY BREEDING

Poultry breeding is done by a wide range of people for diverse end uses and purposes. Poultry breeding can be divided into three main areas which include:

- Commercial breeding.
  - For egg production.
  - Meat production.
- Village/Backyard breeding.
  - Poultry bred for both eggs and meat on a small scale.
- Fancy/Exhibition breeding.
  - Non commercial production of small poultry breeds by enthusiasts.

Commercial Breeding

Chicken is by far the most popular poultry species utilised by Australians for both meat and egg production. Breeding for the commercial poultry sector is on a large industrial scale and hatcheries supply both the broiler and layer industries. In Australia, the term "broiler" or "meat chicken" is used by the industry to describe a chicken grown for meat, while the term "layer" is used for chickens grown and maintained for egg production. Chickens are also affectionately referred to as "chooks". Other poultry species such as Turkeys, Ducks & Geese, and game birds such as Quails are also produced in Australia for meat. Emus and Ostriches are also bred for commercial purposes.

Village/Backyard Breeding

Many farmers and some suburban householders still like to keep their own poultry for egg and meat production. Most of them buy commercial crossbred hens at the point of lay and keep them in semi-intensive conditions in the yard. Some use small colony cages or even use a few layer cages in a protected spot. Others buy day-old chicks and rear their own birds.

Fancy/Exhibition Breeding

There are many breeds of poultry which play very little part in the commercial poultry industry. These are called fancy poultry and are usually kept by small producers (or fanciers) who enjoy breeding, showing and exchanging birds with other fanciers.

COMMERCIAL POULTRY BREEDING

The poultry industry breeds chickens destined for both commercial egg and meat production. Geneticists design special breeding programs to select birds with the best characteristics for egg or meat production. This selection process (called genetic selection or genetics) allows the industry to select strains of birds which are produced very efficiently in intensive housing systems. There are two main types of commercial chicken breeds: layers and meat (broiler) chickens.

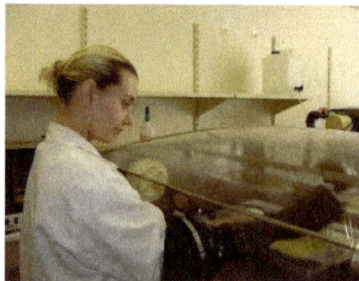

Scientific research is a fundamental component of the poultry industry.

Forming an Egg

The hen releases a yolk with the egg cell in it from her ovary where it moves into the oviduct (egg production tube). When a cockerel and a hen are mated, the sperm cells from the cockerel fertilise the egg cell at the top of the oviduct (fertilisation is the joining of the female egg cell with the male sperm cell). The fertilised egg yolk then takes 23-26 hours to pass down the oviduct, during which time layers of egg white (albumen) are laid down. Two layers of egg membranes are then overlaid, and finally the egg shell. If an egg is not fertilised, it still goes through the same process in the oviduct but it will not develop into a chick.

The Egg

Although the surface of the egg is covered with bacteria, it has its own protective mechanisms in place to prevent the bacteria spoiling the egg. These are:

- The egg cools off after it is laid and bacteria are less able to grow at lower temperatures.

- The shell is coated with a fine moist layer called the cuticle, which dries and protects the egg contents from invading bacteria. This also gives a pleasing appearance, or bloom, to the fresh egg.

Most eggs are laid in the morning. Eggs are collected as soon as possible after laying and placed in a cool room to help preserve their internal quality. Fertile eggs can be stored for up to 7 days at about 12-15oC without loss of hatchability. Because of the danger of bacteria on egg shells going to the hatchery, all fertile eggs are fumigated on the farm or as soon as they arrive at the hatchery. Fumigation with the gas formaldehyde kills surface bacteria without damaging the fertilised ovum inside the egg.

Hatchery

The hatchery is a special building with controlled ventilation. It contains machines for holding and incubating large number of eggs. The hatchery is designed with hygiene in mind and is laid out so that there is little chance of any contaminating organisms travelling back from hatched chicks to eggs brought in later.

Modern hatcheries use specialised equipment.

Stages of Incubation

- First Stage of Incubation:

The first stage lasts for 18 days and is called "setting". During setting, the eggs are placed on special trays which can be tilted through 90 degrees, from side to side. The temperature and humidity of the air in the setter is controlled so that conditions inside each egg are suitable for the growth and development of the chick.

- Second Stage of Incubation:

On the 18th day, eggs are transferred to a different tray, which cannot be tilted, and placed in another machine called a "hatcher". Eggs are transferred to hatchers so that hatching chicks do not contaminate other batches of eggs being incubated. The hatchers can then be thoroughly cleaned after every hatch. By the end of the 21st day all chicks have hatched and are ready to be removed from the machine. They are taken to a special room and removed from the hatcher tray. They are then placed in chick boxes (usually up to 100 in a box) ready for delivery to a farm.

Candling of Eggs

Candling of eggs is done after 5-8 days of incubation to examine for the presence of any infertile eggs. It is the easiest way to check on the development of the chicks inside the eggs.

Hatching chick.

Chick Sexing

Sexing allows separation of male and female chicks. This can be done by:

- Visual examination, (called vent sexing) either by checking the structures in the chick's vent with the naked eye or by inspecting the internal sexual organs with a special lamp.

- Most breeds can now be sexed by checking the feather colour or the degree of growth of wing feathers. Genetic selection has been carried out to ensure that these differences between sexes are distinctive.

Layer chicks are always sexed, as the females are kept while the males are killed. Breeders are usually sexed, as a greater number of females to males are kept for breeding purposes. Meat chickens are normally left unsexed, as both sexes are usually reared together.

Other Procedures

Some vaccines can be administered in the hatchery. In meat chickens, all the required vaccines are administered in the hatchery before delivery to a farm. Beak trimming is sometimes carried out and in some breeds, the comb of the cockerels is trimmed (called dubbing). These procedures may seem cruel but they are carried out to prevent further injury later in life. Beak trimming is done in layer chicks to prevent pecking other birds, or cannibalism. Dubbing is done to prevent injuries to the comb which can result from fighting.

Day old chicks.

Chick Requirements

The baby chick must be kept warm as it does not have the ability to maintain a constant body temperature. The chicks are transported in chick boxes which are designed to conserve heat while allowing air movement. The room where chicks are held in the hatchery and the truck which delivers them to the farm are also designed to keep the chicks both warm (32-34 °C) and ventilated. There is enough food and water in the yolk to keep the chicks alive for about three days, but best results are obtained if they can eat and drink as soon as possible. When placed on the farm, they must be kept warm and have feed and water available at all times.

POULTRY FARM EQUIPMENTS

Incubation Equipments

Setter

- It is a machine in which proper temperature, humidity and turning are provided for the first 19 days of incubating chicken egg.

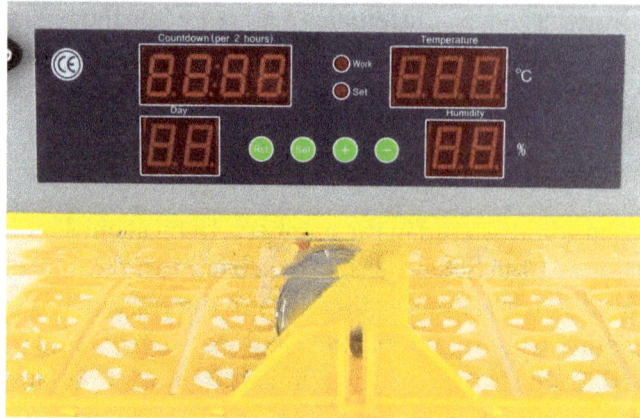

Setter

Hatcher

Walk-in incubators.

Tunnel type incubators.

- It is similar to that of setter but turning mechanism is not available and the trays are designed to hold the newly hatched chicks.

- Here, the eggs are placed for the last three days of incubation.

- Various styles of setter and hatcher found around the world include:

  ○ Walk-in or Corridor incubators.

  ○ Tunnel type incubators.

  ○ Vertical fan incubators.

Compressed Air System

- Some incubators require compressed air to actuate the turning mechanism for the racks of eggs.

- A large central compressed air system is needed for blowing down dust and other dry cleaning in the hatchery.

Emergency Standby Electric Plants

Emergency standby electric plants.

- When there is a failure in the local electrical supply, the incubators must have a secondary source of electricity.

- Therefore, a standby electrical generator located on site, generally within, or next to the hatchery building is imperative.

- The standby electrical generator should have the capacity to support the all essential services of the hatchery.

Hatchery Automation Equipments

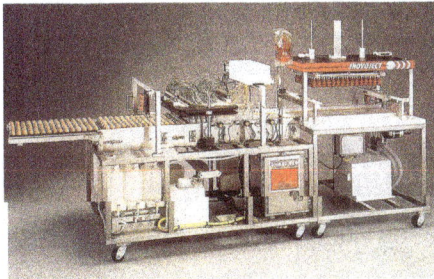

In ovo vaccination.

Hatcher tray washers.

- Hatcher tray washers,

- Waste removal systems,

- Egg transfer machines,

- In ovo vaccination equipment,

- Chick box washers,

- Rack washers,

- Vaccinating/sexing/Grading systems,

- High pressure pumps.

Egg Handling Equipments

Hatching Egg Trays

Vacuum egg lifter.

- This is used to transfer the eggs from the breeder farm trays to hatcher trays.

- Vacuum egg lifts usually employed in the hatcheries handling large volume of eggs.

Egg Candler

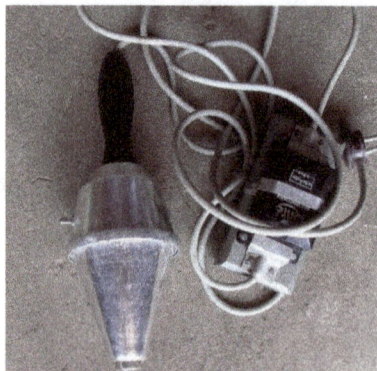

Individual Candler.

- It is a lighting device, used to find out the internal structure of eggs.

- Two types of egg candlers are available, individual and mass candlers.

Brooder Equipments

- Equipment used to provide warmth and light and to rear the baby chicks during the first few weeks of life are called brooders.

- The brooders consist of some heating source, reflectors to reflect the heat and light towards the chicks, light and heat adjustment devices such as stands, thermostats and other accessories, depending on the model.

Charcoal Stove/Kerosene Stove

Charcoal stove.

- These are used in places where electricity is not available or costly and where power failure is quite common.

- These stoves are covered with plates or pans to sustain the heat in the brooding area.

Gas Brooder

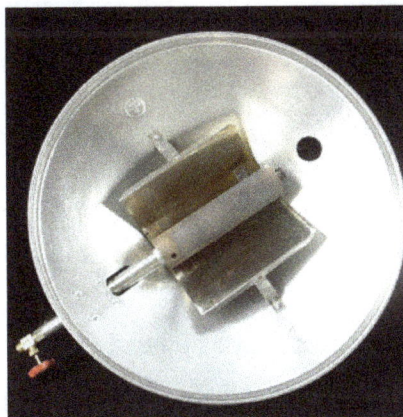

Gas brooder.

- Natural gas, LPG or methane is connected to heating element which is hanged 3 to 5 feet above the chick to provide heat.

- It is attached with canopy type reflectors to reflect the heat towards the chicks.

Electrical Brooder

Electrical brooder.

- It is also thermostatically controlled heating system that spread required amount of heat uniformly above large area, this avoid crowding of chicks under brooder directly.

- One electrical brooder can be used for 300 to 400 chicks.

Infra-red Bulbs

Infra-red bulbs.

- It is a self reflecting bulb and hence no need of reflector over the bulbs.

- 150 and 250 watt bulbs are available to provide sufficient heat to 150 and 250 chicks, respectively.

Reflectors/Hovers

Reflectors/Hovers.

- These reflectors are called Hovers.

- These are reflectors of heat and light.

Flat Type Hover

- These hovers are flat provided with heating element, heating mechanism and pilot lamp and in some cases thermometer are also there in order to record the temperature.

- Generally they are mounted with stands on all four corners, instead of hanging from the roof.

Canopy Type Hover

- These reflectors are in concave shape consisting of ordinary electrical bulb, thermostat mechanism and in some cases thermometer.

Brooder Guard/Chick Guard

Brooder guard.

- These are thin sheets of metal, hard board, or bamboo mat of 1 to 1.5 feet height and varying in lengths.

- They are used to restrict the movement of chicks, so that the chicks will be kept closer to the brooders and prevent them from chilling.

- They are used to prevent chicks from straying too far away from heat supply until they learn the source of heat.

- We have to provide brooder guard with a diameter of 5 feet, height of the brooder should not exceed 1.5 feet.

- For this purpose, we can use materials like cardboard sheet, GI sheet, wire mesh, and mat etc. depending upon the season of brooding.

- During summer season, brooding is done for 5-6 days. In winter season it is 2-3 weeks.

Electrical Heaters (Heating Rods or Coils)

Electrical brooding heaters.

- This type of brooder is provided with heating elements and pilot lamps and in some cases thermometer is provided to record the temperature.

- They used to have a reflecting device over the heating rods or coils.

- The temperature can be adjusted depending on the requirement.

Feeding Equipments

- Feeders are equipment used to feed the birds, by placing feed in them.

- They may be conventional, semi-automatic of various designs and shapes and made up of either metal or plastic.

Automatic Feeder

- In case of automatic feeder the feed is supplied to the entire length of the poultry house by specially designed feed troughs with auger type or chain type devices to move the feed from the feed bins to the other end.

Automatic Feeder.

- These are operated with electricity and the height of the feeder can be adjusted depending upon the age of the birds.

Linear Feeder

Linear Feeder.

- Different sizes of linear feeder with guards are available.

- Provision is also made to adjust the height of the feeder.

- Linear feeders are usually made of Galvanized Iron. However it can as well be made out of any locally available material like wood, bamboo, etc.

- Provisions for stability and adjustment in height at which the feeder stands have to be made in its design.

- Birds can stand on either side of the linear feeder.

- Total feeder space available = 2* length.

- No of linear feeders = (2*Length of the feeder) ÷ Feeder space with all measurements in cm.

Circular Feeder

Circular Feeder.

- These are semi-automatic feeders and can hold 5 to 7 kg feed in its cone at a time.

- The feed is slowly delivered to the bottom by gravity.

- It can also be attached with feed grills to prevent wastage.

- These are made of high plastic and usually suspended from roof/roof-truss or from separate pipeline for the purpose.

- These are also called as 'hanging feeders'.

- These feeders are available in different capacity and when completely full, the feed will suffice 4 to 7 days, depending upon the age and number feeding on them.

- The height at which the feed is available can be easily adjusted by simple clamp mechanism.

- Plastic feeders will be brightly colored (red or blue, generally) and hence are expected to attract layers, especially chicks to feed.

- No. of hanging feeders = 1.3* (Circumference ÷ Feeder space) with all measurements in cm.

- 30% more birds can be accommodated in a hanging feeder when compared to that in linear feeder.

Shell Grit Box

- It is used to provide shell grit to the layer birds as a supplemental source of calcium.

Shell grit box.

Water Equipments

Water Softeners and Filters

- Water with high total dissolved solids will cause deposits on the humidity controls, spray nozzles, jets and valve seats.

- Therefore filter systems and water softeners are necessary to reduce the TDS content of the water used for hatchery operations.

Water Heaters

Water heaters.

- Hot water will be necessary for operating most hatchery tray washers and for general clean up.

- A large capacity boiler is generally used to provide hot water.

Watering Equipments

- Waterers or drinkers are used to provide water to the birds.

- Waterers are available in different sizes, design and shape.

Pan and Jar Type

Pan and Jar type.

- This type of waterer is circular in nature, having two compartments i.e. jar for filling water and pan for delivering water.

Linear Waterer/Channel Type Waterers

- This type waterer is usually attached with cages for providing continuous water supply.

- One end of channel type waterer is designed as funnel shape to receive water from a tap and the other end has the provision for draining the excess water.

Water Basin made of Plastic/Wood/GI with Grill

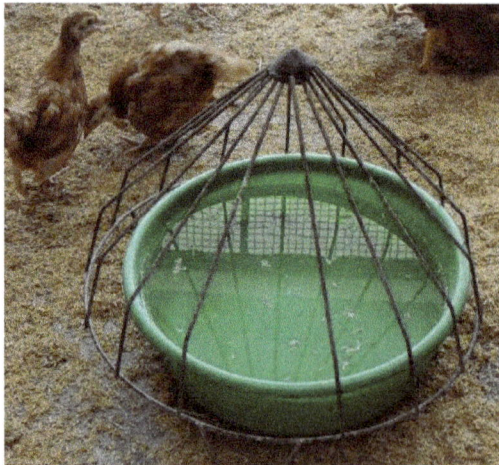
Water basin made of plastic.

- Basins of different diameters are available (10", 12", 14" and 16" diameter).

- A separate grill is available to prevent the entry of birds inside the water.

Bell Type Automatic Waterer

Bell type automatic waterer.

- These are made of high-impact plastic in a bell shape usually suspended from separate pipeline for the purpose.

- This type of waterers has control over the water flow and maintains the required water level always.

- There will be a continuous flow of water so as to ensure water available for the birds throughout the day.

- Height at which the water is available can be easily adjusted by simple clamp mechanism and rate of flow water is adjustable by a valve (spring-mounted). Plastic drinkers will be brightly colored (red,blue) and hence are expected to attract layers, especially chicks to water.

- No. of bell-drinkers=1.3*(circumference÷ Drinker space).

Nipple Drinker

Nipple drinker.

- It can be used both in deep-litter and in cage system.

- When used in deep-litter system, it is attached with cup under the nipple to prevent wetting of litter material.

- These drinkers look like a nipple and water drops comes out when they are pressed.

- They can be used for all types and classes of birds, but most commonly used in laying cages.

- One nipple drinkers in each cage housing 3 layers is sufficient.

Manual Drinker

Manual drinker.

- In case of chicks during first week of brooding, manual drinkers are popularly used.

- They also referred as "fountain drinkers" because water comes out of the holes like that in case of a fountain.

- The main advantage of manual drinkers is the ease of giving vitamins and other probiotics/medicines/vaccines through water.

- Manual drinkers with stand made of high-impact plastic in bright colors (red or blue) are available.

- Arrangement of drinkers at an equal distance of 0.6m between any two feeders and feeder and a drinker.

Vaccination Equipments

Syringe with Needle/Vaccine Droppers

- It is used to give vaccine drops through intra-nasal or intra-ocular.

Automatic Vaccinator

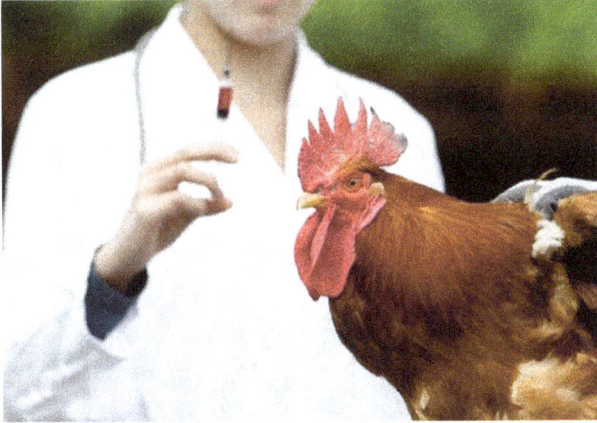

Automatic vaccinator.

- It is used to inject different doses of vaccine to large number of birds in shorter period either through intra-muscular or subcutaneous route.

Fowl Pox Vaccinator/Lancet

- These are used to give fowl pox vaccine at intra-dermal route in the wing web region.

Miscellaneous Equipments

Beak Trimmer

Beak trimmer.

- It is an electrical device used to cut a portion of beak in order to prevent cannibalism among birds.

- The equipment will be mounted on to a stand of convenient height (0.60 to 0.75) with a peddle connected to the top of the unit with a chain/strong thread so that upon pressing the peddle with the foot of the operator, the hot blade slides down cutting the beak placed over a small platform in the equipment.

- The equipment is also provided with a thermostat to regulate temperature.

Nest Boxes

Nest boxes.

- These are used to get clean eggs and to avoid floor eggs in layer or breeder houses.

- These may be individual, communal or trap nest.

Weighing Bala

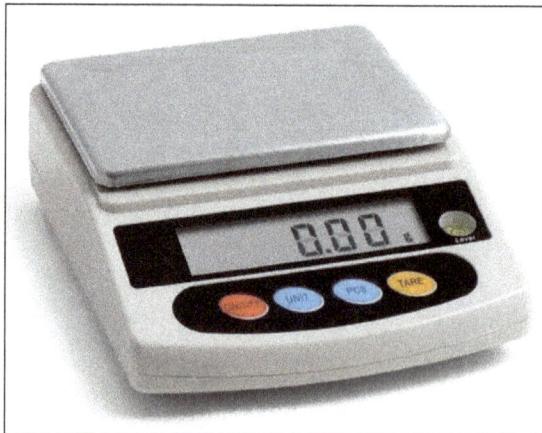

Weighing balances.

- Different types of weighing balances are available to weigh birds or feed for record and marketing purposes.

Perches/Roost

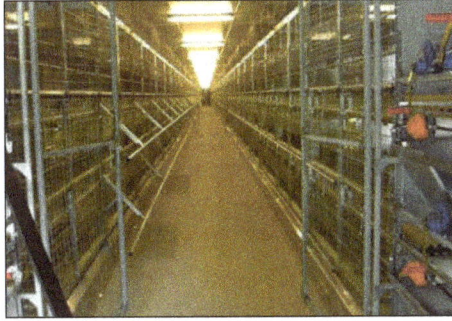

Perches Roost.

- This is a wooden device usually kept at a height of about 3-5' from the floor in order to help the bird to stand over it.

Rake

Rake.

- It is made up of iron rods and fitted with wooden handle.

- This is used to rake the litter material in case of deep-litter system of rearing.

Sprinkler

Sprinkler.

- This is particularly important in hot weather areas.

- Commercial irrigation sprinklers can be used to cool not only the surroundings of the farm buildings but also the roof of the farm.

- But under hot-humid conditions, sprinklers are used only to cool the roof during afternoon hours only.

Sprayer

- Several types of sprayers are available in the market.

- But, that which is hand-operated and can be carried on the back while in use is most ideal for a poultry farm.

- The desired disinfectant or sanitizer can be mixed and loaded on to the tank and sprayed.

Flame-gun (Blow-lamp)

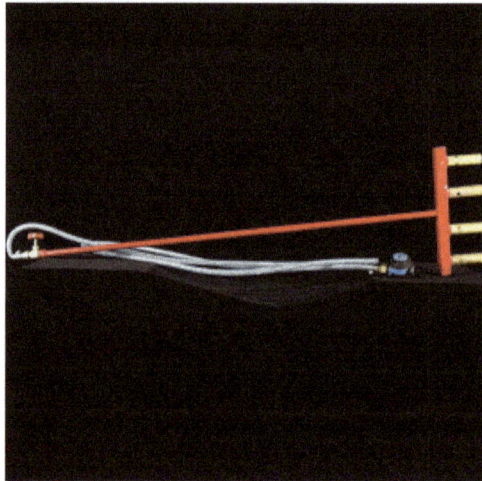

Flame-gun

- It is very useful equipment and it generally works on kerosene (or gas).

- It is used to flame metal frames to rid the building from external parasites and/or their eggs/larva etc.

References

- Poultry-Science, Animal-Science, category, directory: study.com, Retrieved 19 March, 2019

- Semeyn, E. (2002). "Rheobatrachus silus". Animal Diversity Web. University of Michigan Museum of Zoology. Retrieved 2012-08-05

- Incubation%20and%20Hatching, poultry, expert-system: agritech.tnau.ac.in, Retrieved 20 April, 2019

- Poultry-litter-a-great-resource-to-utilize: permaculturenews.org, Retrieved 21 May, 2019

- Products-processing, poultry-production-products: fao.org, Retrieved 22 June, 2019

- Poultry-processing, technology: britannica.com, Retrieved 23 July, 2019

- De Marchi, G., Chiozzi, G., Fasola, M. (2008). "Solar incubation cuts down parental care in a burrow nesting tropical shorebird, the crab plover Dromas ardeola". Journal of Avian Biology 39 (5): 484–486

- Poultry-breeding, industry-structure-and-organisations, production: poultryhub.org, Retrieved 24 August, 2019

- Commercial-poultry-breeding, poultry-breeding, industry-structure-and-organisations, production: poultryhub.org, Retrieved 25 January, 2019

- Poultry%20Farm%20Equipments, poultry, expert-system: agritech.tnau.ac.in, Retrieved 26 February, 2019

2

Types of Poultry

There are numerous birds which fall under the category of poultry. Some of them are chicken, ostriches, goose, domestic ducks, domestic pigeon, squabs, peafowls, domestic turkeys and guinea fowls. The topics elaborated in this chapter will help in gaining a better perspective about these types of poultry.

CHICKEN

The chicken (Gallus gallus) is one of humankind's most common and wide-spread domestic animals. The chicken is believed to be descended from the wild Indian and south-east Asian red junglefowl (also Gallus gallus). They are members of the Phasianidae, or pheasant, family of birds.

Chickens benefit humans greatly as a source of food, both from their meat and their eggs. People in many cultures have admired the good qualities of chickens and have worked to create useful and beautiful breeds. The chicken also has played roles in Chinese religion, Hindu ceremonies, ancient Greek mythology, ancient Roman oracles, Central European folk tales, and in traditional Jewish practice, and are referred to Biblical passages. On the other hand, chickens have often been the victims of human cruelty, for instance in the sport of cockfighting and the inhumane practices in modern factory farms.

A Rooster (male chicken).

Chicken Ancestry: The Red Junglefowl

Red junglefowl.

The range of the red junglefowl stretches from northeast India eastwards across southern China and down into Malaysia and Indonesia. They are about the size of the smaller breeds of domestic chicken, weighing about 0.5 to 1 kilograms (1 to 2 pounds).

Male and female birds show very strong sexual dimorphism. Male junglefowl are larger and they have large red fleshy wattles on the head. The long, bright gold, and bronze feathers form a "shawl" or "cape" over the back of the bird from the neck to the lower back. The tail is composed of long, arching feathers that initially look black but shimmer with blue, purple, and green in good light. The female's plumage is typical of this family of birds in being cryptic and designed for camouflage as she looks after the eggs and chicks. She also has no fleshy wattles or comb on her head.

Junglefowl live in small groups. As in other members of the pheasant family, newly-hatched junglefowl chicks are fully feathered and are able to walk and find food for themselves. The mother hen watches over the chicks and leads them to feeding areas. The roosters seem to play a role in watching over the flock and warning the others of danger.

During the breeding season, the male birds announce their presence with the well known "cock-a-doodle-doo" call. This serves both to attract potential mates and to make other male birds in the area aware of the risk of fighting a breeding competitor. The lower leg just behind and above the foot has a long spur for just this purpose. Their call structure is complex and they have distinctive alarm calls for aerial and ground predators to which others react appropriately.

Flight in these birds is almost purely confined to reaching their roosting areas at sunset in trees or any other high and relatively safe places free from ground predators, and for escape from immediate danger through the day. They feed on the ground, eating mainly seeds and insects.

Domestication

The red junglefowl was probably first domesticated in India around 3000 B.C.E. It is thought that they were first kept as pets rather than as a source of food, although both the birds and their eggs were eaten. Fights were staged between roosters and cockfighting became a popular form of entertainment; it remained so until modern times when these type of bloodsports were banned in many countries.

A day-old chick.

Domestic chickens spread from India east to China about 1400 B.C.E. and west to Egypt about the same time. They entered Europe by way of Persia and Greece soon after. They seem to have been introduced to South America either by Polynesian or Chinese visitors and were later introduced to the rest of the world by European colonists.

Domesticated chickens differ from wild junglefowl in several features. They are usually larger. They are much less nervous and afraid of humans.

Distinct breeds of chickens arose in different locations. In most places, the ability of the rooster to fight was the most important feature chicken breeders selected for, while in both China and ancient Rome chicken meat became important as food and larger breeds were developed. The Chinese developed fancy breeds with beautiful and unusual plumage, while the Romans breed white chickens in order to sacrifice them to their gods.

Names

Male chickens are known as roosters in the United States, Canada, and Australia; in the United Kingdom they are known as cocks when over one year of age, or cockerels when under one year of age. Castrated roosters are called capons. Female chickens over a year old are known as hens. Young females under a year old are known as pullets. Roosters can usually be differentiated from hens by their striking plumage, marked by long flowing tails and bright pointed feathers on their necks. Baby chickens are call chicks.

Behavior

Domestic chickens are not capable of long distance flight, although they are generally capable of flying for short distances such as over fences. Chickens will sometimes fly to explore their surroundings, but usually do so only to flee perceived danger. Because of the risk of escape, chickens raised in open-air pens generally have one of their wings clipped by the breeder—the tips of the longest feathers on one of the wings are cut, resulting in unbalanced flight, which the bird cannot sustain for more than a few meters.

Rooster crowing during daylight hours.

Chickens often Scratch at the Soil to Search for Insects and Seeds. Chickens are gregarious birds and live together as a flock. They have a communal approach to the incubation of eggs and raising of young. Individual chickens in a flock will dominate others, establishing a "pecking order," with dominant individuals having priority for access to food and nesting locations. In the wild, this helps to keep order in the flock, while in domestication it can often lead to injuries or death.

Removing hens or roosters from a flock causes a temporary disruption to this social order until a new pecking order is established. Incidents of cannibalism can occur when a curious bird pecks at a preexisting wound or during fighting (even among female birds). This is exacerbated in close quarters. In commercial egg and meat production, this is controlled by trimming the beak (removal of two thirds of the top half and occasionally one third of the lower half of the beak).

Chickens will try to lay in nests that already contain eggs, and have been known to move eggs from neighboring nests into their own. The result of this behavior is that a flock will use only a few preferred locations, rather than having a different nest for every bird. Some farmers use fake eggs made from plastic or stone to encourage hens to lay in a particular location.

Hens can be extremely stubborn about always laying in the same location. It is not unknown for two (or more) hens to try to share the same nest at the same time. If the nest is small, or one of the hens is particularly determined, this may result in chickens trying to lay on top of each other.

Contrary to popular belief, roosters do not crow only at dawn, but may crow at any time of the day or night. Their crowing—a loud and sometimes shrill call—is a territorial signal to other roosters. However, crowing may also result from sudden disturbances within their surroundings.

When a rooster finds food, he may call the other chickens to eat it first. He does this by clucking in a high pitch as well as picking up and dropping the food. This behavior can also be observed in mother hens, calling their chicks.

In some cases, the rooster will drag the wing opposite the hen on the ground, while circling her. This is part of chicken courting ritual. When a hen is used to coming to his "call" the rooster may mount the hen and proceed with the fertilization.

Chicken eggs vary in color depending on the hen, typically ranging from bright white to shades of brown and even blue, green, and recently reported purple (found in South Asia) (Araucana varieties).

Sometimes a hen will stop laying and instead will focus on the incubation of eggs, a state that is commonly known as going broody. A broody chicken will sit fast on the nest, and protest or peck in defense if disturbed or removed, and will rarely leave the nest to eat, drink, or dust bathe. While brooding, the hen maintains constant temperature and humidity, as well as turning the eggs regularly.

At the end of the incubation period, which is an average of 21 days, the eggs (if fertilized) will hatch, and the broody hen will take care of her young. Since individual eggs do not all hatch at exactly the same time (the chicken can only lay one egg approximately every 25 hours), the hen will usually stay on the nest for about two days after the first egg hatches. During this time, the newly-hatched chicks live off the egg yolk they absorb just before hatching. The hen can sense the chicks peeping inside the eggs, and will gently cluck to stimulate them to break out of their shells. If the eggs are not fertilized by a rooster and do not hatch, the hen will eventually lose interest and leave the nest.

Modern egg-laying breeds rarely go broody, and those that do often stop part-way through the incubation cycle. Some breeds, such as the Cochin, Cornish, and Silkie, regularly go broody and make excellent maternal figures. Chickens used in this capacity are known as utility chickens.

Chicken Farming

Throughout history, chickens, although very common, have almost always been of secondary importance in farming communities. Small flocks were kept on farms, and chicken meat and eggs were often an important source of family food or extra income.

Free range chickens.

After the fall of the Roman Empire, little attention was paid in the West to chicken breeding until the 1800s when more productive breeds began to be developed. The Leghorn has become the most popular breed for egg production, while Rhode Island Reds, Plymouth Rocks, and some others are the most popular for meat.

On farms in the United States, eggs used to be practically the same as currency, with general stores buying eggs for a stated price per dozen. Egg production peaks in the early spring, when farm expenses are high and income is low. On many farms, the flock was the most important source of income, though this was often not appreciated by the farmers, since the money arrived in many small payments. Eggs were a farm operation where even small children could make a valuable contribution.

The major milestone in twentieth century poultry production was the discovery of vitamin D, which made it possible to keep chickens in confinement year-round. Before this, chickens did not thrive during the winter (due to lack of sunlight), and egg production, incubation, and meat production in the off-season were all very difficult, making poultry a seasonal and expensive proposition. Year-round production lowered costs, especially for broilers.

At the same time, egg production was increased by scientific breeding. Improvements in production and quality were accompanied by lower labor requirements. In the 1930s through the early 1950s, having 1,500 hens was considered to be a full-time job for a farm family. In the late 1950s, egg prices had fallen so dramatically that farmers typically tripled the number of hens they kept, putting three hens into what had been a single-bird cage or converting their floor-confinement houses from a single deck of roosts to triple-decker roosts. Not long after this, prices fell still further and large numbers of egg farmers left the business.

This fall in profitability, accompanied by a general fall in prices to the consumer, resulted in poultry and eggs losing their status as luxury foods. This marked the beginning of the transition from family farms to larger, vertically integrated operations. The vertical integration of the egg and poultry industries was a late development, occurring after all the major technological changes had been in place for years (including the development of modern broiler rearing techniques, the adoption of the Cornish Cross broiler, the use of laying cages, etc).

By the late 1950s, poultry production had changed dramatically. Large farms and packing plants could grow birds by the tens of thousands. Chickens could be sent to slaughterhouses for butchering and processing into prepackaged commercial products to be frozen or shipped fresh to markets or wholesalers. Meat-type chickens currently grow to market weight in six to seven weeks whereas only 50 years ago it took three times as long. This is due to genetic selection and nutritional modifications (and not the use of growth hormones, which are illegal for use in poultry in the United States and many other countries). Once a meat consumed only occasionally, the common availability and lower cost has made chicken a common meat product within developed nations. Growing concerns over the cholesterol content of red meat in the 1980s and 1990s further resulted in increased consumption of chicken.

Modern Chicken Farming

Modern egg production.

Today, eggs are produced on large egg ranches on which environmental parameters are controlled. Chickens are exposed to artificial light cycles to stimulate egg production year-round. In addition, it is a common practice to induce molting through manipulation of light and the amount of food they receive in order to further increase egg size and production.

On average, a chicken lays one egg a day for a number of days (a "clutch"), then does not lay for one or more days, then lays another clutch. Originally, the hen presumably laid one clutch, became broody, and incubated the eggs. Selective breeding over the centuries has produced hens that lay more eggs than they can hatch. Some of this progress was ancient, but most occurred after 1900. In 1900, average egg production was 83 eggs per hen per year. In 2000, it was well over 300.

In the United States, laying hens are butchered after their second egg laying season. In Europe, they are generally butchered after a single season. The laying period begins when the hen is about 18-20 weeks old (depending on breed and season). Males of the egg-type breeds have little commercial value at any age, and all those not used for breeding (roughly fifty percent of all egg-type chickens) are killed soon after hatching. The old hens also have little commercial value. Thus, the main sources of poultry meat 100 years ago (spring chickens and stewing hens) have both been entirely supplanted by meat-type broiler chickens.

Traditionally, chicken production was distributed across the entire agricultural sector. In the twentieth century, it gradually moved closer to major cities to take advantage of lower shipping costs. This had the undesirable side effect of turning the chicken manure from a valuable fertilizer that could be used profitably on local farms to an unwanted byproduct. This trend may be reversing itself due to higher disposal costs on the one hand and higher fertilizer prices on the other, making farm regions attractive once more.

Small Scale and Hobby Chicken Raising

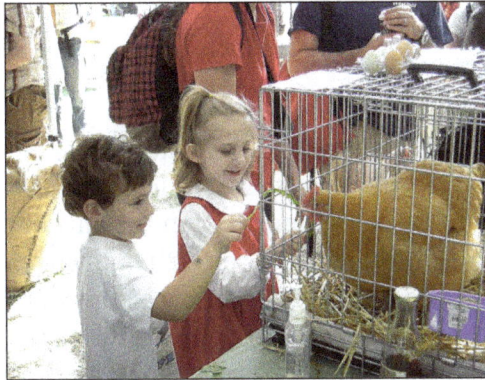

Orpington shown at a fair.

In most of the world, small flocks of chickens are still kept on farms and homesteads as they have been throughout history. In addition some people raise chickens as a hobby or as pets.

Purebred chickens are shown at shows and agricultural fairs. The American Poultry Association recognizes 113 different chicken breeds. Part of their interest is to preserve breeds that are in danger of going extinct because they are no longer being used in factory farming.

Concerns with Modern Chicken Farming

Humane Treatment

Animal welfare groups have frequently targeted the poultry industry for engaging in practices that they believe to be inhumane. Many animal welfare advocates object to killing chickens for food, the "factory farm conditions" under which they are raised, methods of transport, and slaughter. PETA and other groups have repeatedly conducted undercover investigations at chicken farms and slaughterhouses, which they allege confirm their claims of cruelty.

Laying hens are routinely debeaked to prevent fighting. Because beaks are sensitive, trimming them without anesthesia is considered inhumane by some. It is also argued that the procedure causes life-long discomfort. Conditions in intensive chicken farms may be unsanitary, allowing the proliferation of diseases such as salmonella and E. coli. Chickens may be raised in total darkness. Rough handling and crowded transport during various weather conditions and the failure of existing stunning systems to render the birds unconscious before slaughter have also been cited as welfare concerns.

Another animal welfare concern is the use of selective breeding to create heavy, large-breasted birds, which can lead to crippling leg disorders and heart failure for some of the birds. Concerns have been raised that companies growing single varieties of birds for eggs or meat are increasing their susceptibility to disease.

Human Concerns

Antibiotics

Antibiotics have been used on poultry in large quantities since the Forties. This is because it was found that the byproducts of antibiotic production—which were being fed to chickens because of high level of vitamin B12 in the antibiotic-producing mold after removal of the antibiotics—produced higher growth than could be accounted for by just the B12. Eventually it was discovered that the trace amounts of antibiotics remaining in the byproducts accounted for this growth. The mechanism is apparently the adjustment of intestinal flora, favoring "good" bacteria while suppressing "bad" bacteria, and thus the goal of antibiotics as a growth promoter is the same as for probiotics. Because the antibiotics used are not absorbed by the gut, they do not put antibiotics into the meat or eggs.

Antibiotics are used routinely in poultry for this reason, and also to prevent and treat disease. Many contend that this puts humans at risk as bacterial strains develop stronger and stronger resistances. Critics of this view point out that, after six decades of heavy agricultural use of antibiotics, opponents of antibiotics must still make arguments about theoretical risks, since actual examples are hard to come by. Those antibiotic-resistant strains of human diseases whose origin is known apparently originated in hospitals rather than farms.

E.Coli

According to Consumer Reports, "1.1 million or more Americans are sickened each year by undercooked, tainted chicken". A USDA study discovered E.Coli in 99 percent of supermarket chicken, the result of chicken butchering not being a sterile process. Feces tend to leak from the carcass until the evisceration stage, and the evisceration stage itself gives an opportunity for the interior of the carcass to receive intestinal bacteria. (So does the skin of the carcass, but the skin presents a better barrier to bacteria and reaches higher temperatures during cooking).

A free roaming bantam chickens.

Before 1950, this bacterial risk was contained largely by not eviscerating the carcass at the time of butchering, deferring this until the time of retail sale or in the home. This gave the intestinal bacteria less opportunity to colonize the edible meat. The development of the "ready-to-cook broiler" in the 1950s added convenience while introducing risk, under the assumption that end-to-end refrigeration and thorough cooking would provide adequate protection.

E. Coli can be killed by proper cooking times, but there is still some risk associated with it, and its near-ubiquity in commercially-farmed chicken is troubling to some. Irradiation has been proposed as a means of sterilizing chicken meat after butchering; while proper storage, handling, and cooking are always important.

Avian influenza

There is also a risk that the crowded conditions in many chicken farms will allow avian influenza to spread quickly. A United Nations press release states: "Governments, local authorities and international agencies need to take a greatly increased role in combating the role of factory-farming, commerce in live poultry, and wildlife markets which provide ideal conditions for the virus to spread and mutate into a more dangerous form".

Efficiency

Farming of chickens on an industrial scale relies largely on high protein feeds derived from soybeans; in the European Union the soybean dominates the protein supply for animal feed, and the poultry industry is the largest consumer of such feed. Giving the feed to chickens means the protein reaches humans with a much lower efficiency than through direct consumption of soybean products. Some nutrients, however, are present in chicken but not in the soybean.

OSTRICH

Common ostrich (*Struthio camelus*), male and female.

The ostriches are a family, Struthionidae, of flightless birds. The two extant species of ostrich are the common ostrich and Somali ostrich, both in the genus *Struthio*, which also contains several species known from Holocene fossils such as the Asian ostrich. The common ostrich is the more widespread of the two living species, and is the largest living bird species. Other ostriches are also among the largest bird species ever.

Ostriches first appeared during the Miocene epoch, though various Paleocene, Eocene, and Oligocene fossils may also belong to the family. Ostriches are classified in the ratite group of birds, all extant species of which are flightless, including the kiwis, emus, and rheas. Traditionally, the order Struthioniformes contained all the ratites. However, recent genetic analysis has found that the group is not monophyletic, as it is paraphyletic with respect to the tinamous, so the ostriches are classified as the only members of the order.

The earliest fossils of ostrich-like birds are Paleocene taxa from Europe. *Palaeotis* and *Remiornis* from the Middle Eocene and unspecified ratite remains are known from the Eocene and Oligocene of Europe and Africa. These may have been early relatives of the ostriches, but their status is questionable, and they may in fact represent multiple lineages of flightless paleognaths. The African *Eremopezus*, when not considered a basal secretarybird or shoebill, is sometimes considered an ostrich relative or an "aepyornithid-like" taxon. Apart from these enigmatic birds, the fossil record of the ostriches continues with several species of the modern genus *Struthio*, which are known from the Early Miocene onwards. Several of these fossil forms are ichnotaxa (that is, classified according to the organism's footprints or other trace rather than its body) and their association with those described from distinctive bones is contentious and in need of revision pending more good material. While the relationship of the African fossil species is comparatively straightforward, a large number of Asian species of ostriches have been described from fragmentary remains, and their interrelationships and how they relate to the African ostriches are confusing. In China, ostriches are known to have become extinct only around or even after the end of the last ice age; images of ostriches have been found there on prehistoric pottery and petroglyphs.

Ostriches have co-existed with another lineage of flightless didactyl birds, the eogruids. Though Olson 1985 classified these birds as stem-ostriches, they are otherwise universally considered to be related to cranes, any similarities being the result of convergent evolution. Competition from ostriches has been suggested to have caused the extinction of the eogruids, though this has never been tested and both groups do co-exist in some sites.

Distribution and Habitat

A male Somali ostrich in a Kenyan savanna, showing its blueish neck.

Today ostriches are only found natively in the wild in Africa, where they occur in a range of open arid and semi-arid habitats such as savannas and the Sahel, both north and south of the equatorial forest zone. The Somali ostrich occurs in the Horn of Africa, having evolved isolated from the common ostrich by the geographic barrier of the East African Rift. In some areas, the common ostrich's Masai subspecies occurs alongside the Somali ostrich, but they are kept from interbreeding by behavioral and ecological differences. The Arabian ostriches in Asia Minor and Arabia were hunted to extinction by the middle of the 20th century, and in Israel attempts to introduce North African ostriches to fill their ecological role have failed. Escaped common ostriches in Australia have established feral populations.

BROILER

A broiler is any chicken (*Gallus gallus domesticus*) that is bred and raised specifically for meat production. Most commercial broilers reach slaughter weight between four and seven weeks of age, although slower growing breeds reach slaughter weight at approximately 14 weeks of age. Typical broilers have white feathers and yellowish skin.

Due to extensive breeding selection for rapid early growth and the husbandry used to sustain this, broilers are susceptible to several welfare concerns, particularly skeletal malformation and dysfunction, skin and eye lesions and congestive heart conditions. Management of ventilation, housing, stocking density and in-house procedures must be evaluated regularly to support good welfare of the flock. The breeding stock (broiler-breeders) do grow to maturity but also have their own welfare concerns related to the frustration of a high feeding motivation and beak trimming. Broilers are usually grown as mixed-sex flocks in large sheds under intensive conditions.

Modern Breeding

Before the development of modern commercial meat breeds, broilers were mostly young male chickens culled from farm flocks. Pedigree breeding began around 1916. Magazines for the poultry industry existed at this time. A crossbred variety of chicken was produced from a male of a naturally double-breasted Cornish strain, and a female of a tall, large-boned strain of white Plymouth Rocks. This first attempt at a meat crossbreed was introduced in the 1930s and became dominant in the 1960s. The original crossbreed was plagued by problems of low fertility, slow growth and disease susceptibility.

Modern broilers have become very different from the Cornish/Rock crossbreed. As an example, Donald Shaver (originally a breeder of egg-production breeds) began gathering breeding stock for a broiler program in 1950. Besides the breeds normally favoured, Cornish Game, Plymouth Rock, New Hampshire, Langshans, Jersey Black Giant and Brahmas were included. A white feathered female line was purchased from Cobb. A full-scale breeding program was commenced in 1958, with commercial shipments in Canada and the US in 1959 and in Europe in 1963.

As a second example, colour sexing broilers was proposed by Shaver in 1973. The genetics were based on the company's breeding plan for egg layers, which had been developed in the mid-1960s. A difficulty facing the breeders of the colour-sexed broiler is that the chicken must be white-feathered by slaughter age. After 12 years, accurate colour sexing without compromising economic traits was achieved.

Artificial Insemination

Artificial insemination is a mechanism in which spermatozoa are deposited into the reproductive tract of a female. Artificial insemination provides a number of benefits relating to reproduction in the poultry industry. Broiler breeds have been selected specifically for growth, causing them to develop large pectoral muscles, which interfere with and reduce natural mating. The amount of sperm produced and deposited in the hen's reproductive tract may be limited because of this. Additionally, the males' overall sex drive may be significantly reduced due to growth selection. Artificial insemination has allowed many farmers to incorporate selected genes into their stock, increasing their genetic quality.

Abdominal massage is the most common method used for semen collection. During this process, the rooster is restrained and the back region located towards the tail and behind the wings is caressed. This is done gently but quickly. Within a short period of time, the male should get an erection of the phallus. Once this occurs, the cloaca is squeezed and semen is collected from the external papilla of the vas deferens.

During artificial insemination, semen is most frequently deposited intra-vaginally by means of a plastic syringe. In order for semen to be deposited here, the vaginal orifice is everted through the cloaca. This is simply done by applying pressure to the abdomen of the hen. The semen-containing instrument is placed 2–4 cm into the vaginal orifice. As the semen is being deposited, the pressure applied to the hen's abdomen is being released simultaneously. The individual performing this procedure typically uses one hand to move and direct the tail feathers, while using the other hand to insert the instrument and semen into the vagina.

General Biology

Modern commercial broilers, for example, Cornish crosses and Cornish-Rocks, are artificially selected and bred for large-scale, efficient meat production. They are noted for having very fast growth rates, a high feed conversion ratio, and low levels of activity. Modern commercial broilers are bred to reach a slaughter-weight of about 2 kg in only 35 to 49 days. As a consequence, the behavior and physiology of broilers reared for meat are those of immature birds, rather than adults. Slow growing free-range and organic strains have been developed which reach slaughter-weight at 12 to 16 weeks of age.

Typical broilers have white feathers and yellowish skin. Recent genetic analysis has revealed that the gene for yellow skin was incorporated into domestic birds through hybridization with the grey junglefowl (*G. sonneratii*). Modern crosses are also favorable for meat production because they lack the typical "hair" which many breeds have that must be removed by singeing after plucking the carcass.

Both male and female broilers are reared for their meat.

Behavior

Broiler behavior is modified by the environment, and alters as the broilers' age and bodyweight rapidly increase. For example, the activity of broilers reared outdoors is initially greater than broilers reared indoors, but from six weeks of age, decreases to comparable levels in all groups. The same study shows that in the outdoors group, surprisingly little use is made of the extra space and facilities such as perches – it was proposed that the main reason for this was leg weakness as 80 per cent of the birds had a detectable gait abnormality at seven weeks of age. There is no evidence of reduced motivation to extend the behavioral repertoire, as, for example, ground pecking remained at significantly higher levels in the outdoor groups because this behavior could also be performed from a lying posture rather than standing.

Examining the frequency of all sexual behavior shows a large decrease with age, suggestive of a decline in libido. The decline in libido is not enough to account for reduced fertility in heavy cocks at 58 weeks and is probably a consequence of the large bulk or the conformation of the males at this age interfering in some way with the transfer of semen during copulations which otherwise look normal.

Feeding and Feed Conversion

Chickens are omnivores and modern broilers are given access to a special diet of high protein feed, usually delivered via an automated feeding system. This is combined with artificial lighting conditions to stimulate eating and growth and thus the desired body weight.

In the U.S., the average feed conversion ratio (FCR) of a broiler was 1.91 pounds of feed per pound of liveweight in 2011, up from 4.70 in 1925. Canada has a typical FCR of 1.72. New Zealand commercial broiler farms have recorded the world's best broiler chicken FCR, consistently at 1.38 or lower.

Welfare Issues

Meat Birds

Artificial selection has led to a great increase in the speed with which broilers develop and reach slaughter-weight. The time required to reach 1.5 kg live-weight decreased from 120 days to 30 days between 1925 and 2005. Selection for fast early growth-rate, and feeding and management procedures to support such growth, have led to various welfare problems in modern broiler strains. Welfare of broilers is of particular concern given the large number of individuals that are produced; for example, the U.S. in 2011 produced approximately 9 billion broiler chickens.

Cardiovascular Dysfunction

Selection and husbandry for very fast growth means there is a genetically induced mismatch between the energy-supplying organs of the broiler and its energy-consuming organs. Rapid growth can lead to metabolic disorders such as sudden death syndrome (SDS) and ascites.

SDS is an acute heart failure disease that affects mainly male fast-growing broilers which appear to be in good condition. Affected birds suddenly start to flap their wings, lose their balance, sometimes cry out and then fall on their backs or sides and die, usually all within a minute. In 1993, U.K. broiler producers reported an incidence of 0.8%. In 2000, SDS has a death rate of 0.1% to 3% in Europe.

Ascites is characterised by hypertrophy and dilatation of the heart, changes in liver function, pulmonary insufficiency, hypoxaemia and accumulation of large amounts of fluid in the abdominal cavity. Ascites develops gradually and the birds suffer for an extended period before they die. In the UK, up to 19 million broilers die in their sheds from heart failure each year.

Skeletal Dysfunction

Breeding for increased breast muscle means that the broilers' centre of gravity has moved forward and their breasts are broader compared with their ancestors, which affects the way they walk and puts additional stresses on their hips and legs. There is a high frequency of skeletal problems in broilers, mainly in the locomotory system, including varus and valgus deformities, osteodystrophy, dyschondroplasia and femoral head necrosis. These leg abnormalities impair the locomotor abilities of the birds, and lame birds spend more time lying and sleeping. The behavioral activities of broilers decrease rapidly from 14 days of age onwards. Reduced locomotion also decreases ossification of the bones and results in skeletal abnormalities; these are reduced when broilers have been exercised under experimental conditions.

Most broilers find walking painful, as indicated by studies using analgesic and anti-inflammatory drugs. In one experiment, healthy birds took 11 seconds to negotiate an obstacle course, whereas lame birds took 34 seconds. After the birds had been treated with carprofen, there was no effect on the speed of the healthy birds, however, the lame birds now took only 18 seconds to negotiate the course, indicating that the pain of lameness is relieved by the drug. In self-selection experiments, lame birds select more drugged feed than non-lame birds leading to the suggestion that leg problems in broilers are painful .

Several research groups have developed "gait scores" (GS) to objectively rank the walking ability and lameness of broilers. In one example of these scales, GS=0 indicates normal walking ability, GS=3 indicates an obvious gait abnormality which affects the bird's ability to move about and GS=5 indicates a bird that cannot walk at all. GS=5 birds tried to use their wings to help them walking, or crawled along on their shanks. In one study, almost 26% of the birds examined were rated as GS=3 or above and can therefore be considered to have suffered from painful lameness.

Young birds being reared in a closed broiler house.

Integument Lesions

Sitting and lying behaviors in fast growing strains increase with age from 75% in the first seven days to 90% at 35 days of age. This increased inactivity is linked with an increase in dermatitis caused by a greater amount of time in contact with ammonia in the litter. This contact dermatitis is characterised by hyperkeratosis and necrosis of the epidermis at the affected sites; it can take forms such as hock burns, breast blisters and foot pad lesions.

Stocking Density

Broilers in a rearing shed indicating the high stocking densities used.

Broilers are usually kept at high stocking densities which vary considerably between countries. Typical stocking densities in Europe range between about 22 to 42 kg/m^2

or between about 11 to 25 birds per square metre. There is a reduction of feed intake and reduced growth rate when stocking density exceeds approximately 30 kg/m² under deep litter conditions. The reduced growth rate is likely due to a reduced capacity to lose heat generated by metabolism. Higher stocking densities are associated with increased dermatitis including food pad lesions, breast blisters and soiled plumage. In a large-scale experiment with commercial farms, it was shown that the management conditions (litter quality, temperature and humidity) were more important than stocking density.

Ocular Dysfunction

In attempts to improve or maintain fast growth, broilers are kept under a range of lighting conditions. These include continuous light (fluorescent and incandescent), continuous darkness, or under dim light; chickens kept under these light conditions develop eye abnormalities such as macrophthalmos, avian glaucoma, ocular enlargement and shallow anterior chambers.

Ammonia

The litter in broiler pens can become highly polluted from the nitrogenous feces of the birds and produce ammonia. Ammonia has been shown to cause increased susceptibility to disease and other health-related problems such as Newcastle disease, airsaculitis and keratoconjunctivitis. The respiratory epithelium in birds is damaged by ammonia concentrations in the air exceeding 75 parts per million (ppm). Ammonia concentrations at 25 to 50 ppm induce eye lesions in broiler chicks after seven days of exposure.

Catching and Transport

Once the broilers have reached the target live-weight, they are caught, usually by hand, and packed live into crates for transport to the slaughterhouse. They are usually deprived of food and water for several hours before catching until slaughter. The process of catching, loading, transport and unloading causes serious stress, injury and even death to a large number of broilers.

The number of broilers that died in the EU in 2005 during the process of catching, packing and transport was estimated to be as high as 18 to 35 million. In the UK, of broilers that were found to be 'dead on arrival' at the slaughterhouse in 2005, it was estimated that up to 40% may have died from thermal stress or suffocation due to crowding on the transporter.

Slaughter is done by hanging the birds fully conscious by their feet upside-down in shackles on a moving chain, stunning them by automatically immersing them in an electrified water bath and exsanguination by cutting their throats.

Some research indicates that chickens might be more intelligent than previously supposed, which "raises questions about how they are treated". A possible 10-year life span has been shortened to six weeks for broilers.

Mortality Rates

According to historical records, broiler mortality rates in the U.S. have decreased from 18% in 1925 to 3.7% in 2012, but have increased since 2013 to reach 5% in 2018.

One indication of the effect of broilers' rapid growth rate on welfare is a comparison of the usual mortality rate for standard broiler chickens (1% per week) with that for slower-growing broiler chickens (0.25% per week) and with young laying hens (0.14% per week); the mortality rate of the fast-growing broilers is seven times the rate of laying hens (the same subspecies) of the same age.

Parent Birds

Meat broilers are usually slaughtered at approximately 35 to 49 days of age, well before they become sexually reproductive at 5 to 6 months of age. However, the bird's parents, often called "broiler-breeders", must live to maturity and beyond so they can be used for breeding. As a consequence, they have additional welfare concerns.

Meat broilers have been artificially selected for an extremely high feeding motivation, but are *not* usually feed-restricted, as this would delay the time taken for them to reach slaughter-weight. Broiler-breeders have the same highly increased feeding motivation, but *must* be feed-restricted to prevent them becoming overweight with all its concomitant life-threatening problems. An experiment on broilers' food intake found that 20% of birds allowed to eat as much as they wanted either died or had to be killed because of severe illness between 11 and 20 weeks of age – either they became so lame they could not stand or they developed cardiovascular problems.

Broiler breeders fed on commercial rations eat only a quarter to a half as much as they would with free access to food. They are highly motivated to eat at all times, presumably leading to chronic frustration of feeding.

Because broiler breeders live to adulthood, they might show feather pecking or other injurious pecking behavior. To avoid this, they might be beak trimmed which can lead to acute or chronic pain.

World Production and Consumption

A report in 2005 stated that around 5.9 billion broiler chickens for eating were produced yearly in the European Union. Mass production of chicken meat is a global industry and at that time, only two or three breeding companies supplied around 90% of the world's breeder-broilers. The total number of meat chickens produced in the world was nearly 47 billion in 2004; of these, approximately 19% were produced in the US, 15% in China, 13% in the EU25 and 11% in Brazil.

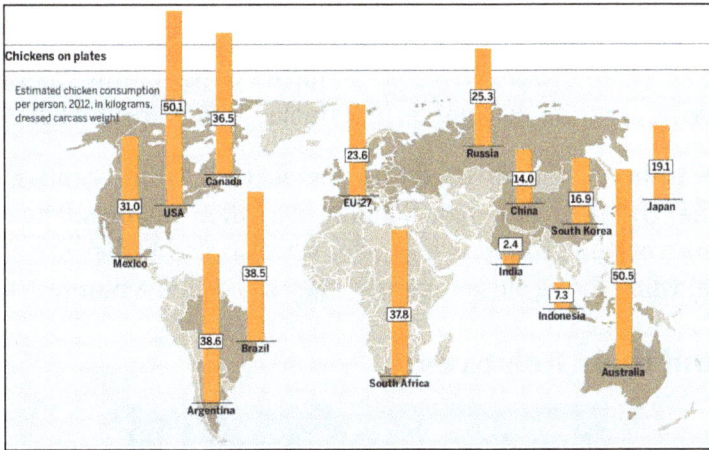

Chickens on plates

Estimated chicken consumption per person, 2012, in kilograms, dressed carcass weight

USA	50.1
Canada	36.5
Mexico	31.0
EU-27	23.6
Russia	25.3
China	14.0
South Korea	16.9
Japan	19.1
India	2.4
Indonesia	7.3
Australia	50.5
Brazil	38.5
Argentina	38.6
South Africa	37.8

Estimated chicken consumption per person in 2012.

Consumption of broilers is surpassing that of beef in industrialized countries. Demand in Asia is rising. Worldwide, 86.6 million tonnes of broiler meat were produced in 2014. As of 2018, the worldwide estimation of broiler chick population is approximately 23 billion.

Broiler Industry

The commercial production of broiler chickens for meat consumption is a highly industrialized process. There are two major sectors:

1. Rearing birds intended for consumption.

2. Rearing parent stock for breeding the meat birds.

GOOSE

A greylag goose (Anser anser).

A goose (plural geese) is a bird of any of several waterfowl species in the family Anatidae. This group comprises the genera *Anser* (the grey geese), *Branta* (the black geese), and *Chen* (which includes the white geese); the latter being commonly placed within the genus *Anser*. Some other birds, mostly related to the shelducks, have "goose" as part of their names. More distantly related members of the family Anatidae are swans, most of which are larger than true geese, and ducks, which are smaller.

The term "goose" is more properly used for a female bird, while "gander" refers specifically to a male one. Young birds before fledging are called goslings. The collective noun for a group of geese on the ground is a gaggle; when in flight, they are called a skein, a team, or a wedge; when flying close together, they are called a plump.

True Geese and their Relatives

Snow geese in Quebec, Canada.

The three living genera of true geese are: *Anser*, grey geese, including the greylag goose, and domestic geese; *Chen*, white geese (often included in *Anser*); and *Branta*, black geese, such as the Canada goose.

Two genera of geese are only tentatively placed in the Anserinae; they may belong to the shelducks or form a subfamily on their own: *Cereopsis*, the Cape Barren goose, and *Cnemiornis*, the prehistoric New Zealand goose. Either these or, more probably, the goose-like Coscoroba swan is the closest living relative of the true geese.

Fossils of true geese are hard to assign to genus; all that can be said is that their fossil record, particularly in North America, is dense and comprehensively documents many different species of true geese that have been around since about 10 million years ago in the Miocene. The aptly named *Anser atavus* (meaning "progenitor goose") from some 12 million years ago had even more plesiomorphies in common with swans. In addition, some goose-like birds are known from subfossil remains found on the Hawaiian Islands.

Geese are monogamous, living in permanent pairs throughout the year; however, unlike most other permanently monogamous animals, they are territorial only during the short nesting season. Paired geese are more dominant and feed more, two factors that result in more young.

Chinese geese, the domesticated form of the swan goose.

Geese

Greylag goose.

Cape Barren goose.

Some mainly Southern Hemisphere birds are called "geese", most of which belong to the shelduck subfamily Tadorninae. These are:

- Orinoco goose, Neochen jubata.

- Egyptian goose, Alopochen aegyptiaca.

- The South American sheldgeese, genus Chloephaga.

- The prehistoric Malagasy sheldgoose, Centrornis majori.

Others:

- The spur-winged goose, *Plectropterus gambensis*, is most closely related to the shelducks, but distinct enough to warrant its own subfamily, the Plectropterinae.

- The blue-winged goose, *Cyanochen cyanopterus*, and the Cape Barren goose, *Cereopsis novaehollandiae*, have disputed affinities. They belong to separate ancient lineages that may ally either to the Tadorninae, Anserinae, or closer to the dabbling ducks (Anatinae).

- The three species of small waterfowl in the genus *Nettapus* are named "pygmy geese". They seem to represent another ancient lineage, with possible affinities to the Cape Barren goose or the spur-winged goose.

- A genus of prehistorically extinct seaducks, *Chendytes*, is sometimes called "diving-geese" due to their large size.

- The unusual magpie goose is in a family of its own, the Anseranatidae.

- The northern gannet, a seabird, is also known as the "Solan goose", although it is a bird unrelated to the true geese, or any other Anseriformes for that matter.

Canada goose gosling.

Canada geese in flight, Great Meadows Wildlife Sanctuary.

DOMESTIC DUCK

Domestic ducks are ducks that are raised for meat, eggs and down. Many ducks are also kept for show, as pets, or for their ornamental value. Almost all varieties of domestic duck apart from the Muscovy duck (*Cairina moschata*) are descended from the mallard (*Anas platyrhynchos*).

A group of Indian Runner ducks.

Domestication

Ducks were common in Ancient Egypt, as depicted in this 3rd century BC faience vase.

Mallard ducks were first domesticated in Southeast Asia at least 4000 years ago, during the Neolithic Age, and were also farmed by the Romans in Europe, and the Malays in Asia. In ancient Egypt, ducks were captured in nets and then bred in captivity. During the Ming Dynasty, the Peking duck—mallards force-fed on grains, making them larger— was known to have good genetic characteristics.

Almost all varieties of domestic duck except the muscovy have been derived from the mallard. Domestication has greatly altered their characteristics. Domestic ducks are mostly polygamous, where wild mallards are monogamous. Domestic ducks have lost the mallard's territorial behavior, and are less aggressive than mallards. Despite these differences, domestic ducks frequently mate with wild mallard, producing fully fertile hybrid offspring.

Farming

A duck farm in Taiwan.

Ducks have been farmed for thousands of years. In the Western world, they are not as popular as the chicken, because chickens have much more white lean meat and are easier to keep confined, making the total price much lower for chicken meat, whereas duck is comparatively expensive. While popular in *haute cuisine*, duck appears less frequently in the mass-market food industry and restaurants in the lower price range. However, ducks are more popular in China and there they are raised extensively.

Ducks are farmed for their meat, eggs, and down. A minority of ducks are also kept for *foie gras* production. The blood of ducks slaughtered for meat is also collected in some regions and is used as an ingredient in many cultures' dishes. Their eggs are blue-green to white, depending on the breed.

Ducks can be kept free range, in cages, in barns, or in batteries. Ducks enjoy access to swimming water, but do not require it to survive. They should be fed a grain and insect diet. It is a popular misconception that ducks should be fed bread; bread has limited nutritional value and can be deadly when fed to developing ducklings. Ducks should be monitored for avian influenza, as they are especially prone to infection with the dangerous H5N1 strain.

The females of many breeds of domestic ducks are unreliable at sitting their eggs and raising their young. Exceptions include the Rouen duck and especially the Muscovy duck. It has been a custom on farms for centuries to put duck eggs under broody hens for hatching; nowadays this role is often played by an incubator. However, young ducklings rely on their mothers for a supply of preen oil to make them waterproof; a chicken hen does not make as much preen oil as a female duck, and an incubator makes none. Once the duckling grows its own feathers, it produces preen oil from the sebaceous gland near the base of its tail.

Pets and Ornamentals

The Muscovy duck (drake shown) is the only domestic breed not derived from the mallard.

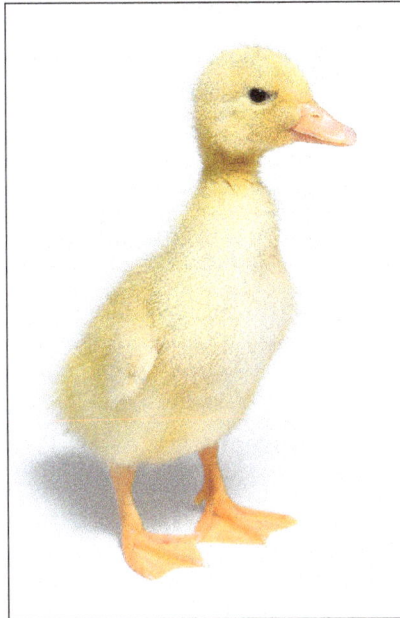

Domestic duckling.

Domestic ducks can be kept as pets, in a garden or backyard, generally with a pond or deep water dish. If they are given access to a pond, they dabble in the mud, dredging out and eating wildlife and frog spawn, and swallow adult frogs and toads up to the size of the common frog *Rana temporaria*, as they have been bred to be much bigger than wild ducks, with a waterline length (base of neck to base of tail) of up to 1 foot (30 cm); the wild mallard's waterline length is about 6 inches (15 cm). Protection from predators such as foxes and hawks is required.

Ducks are also kept for their ornamental value. Breeds have been developed with crests and tufts or striking plumage, for exhibition in competitions.

DOMESTIC PIGEON

Domestic pigeon (*Columba livia*).

Domestic Pigeon is a bird of the family Columbidae (order Columbiformes) that was perhaps the first bird tamed by man. Figurines, mosaics, and coins have portrayed the domestic pigeon since at least 4500 BC (Mesopotamia). From Egyptian times the pigeon has been important as food. Its role as messenger has a long history. Today it is an important laboratory animal, especially in endocrinology and genetics.

Throwbacks among modern domestic pigeons indicate a common ancestor, the rock dove. This tendency is clearly seen in street pigeons in cities everywhere. Many are white, reddish, or checkered like some of their cousins in racing-pigeon lofts, but most are somewhat narrow-bodied and broad-billed replicas of the blue-gray ancestral form. Street pigeons nest year-round, on buildings and beneath bridges, where they may be a nuisance with their droppings and transmission of disease. These hardy birds may live 35 years.

The three main kinds of domestic pigeons are fliers, fancy breeds raised chiefly for show, and utility breeds, which produce squabs for meat—nestlings taken when 25 to 30 days old and weighing 350 to 700 grams (3/4 to 1 1/2 pounds). Utility breeds are known as dual-purpose birds if they are bred to exhibition standards.

Pigeon-raising is a worldwide hobby, and business as well. National preferences are evident; e.g., in England for birds of highly standardized appearance and bearing ("form pigeons"), in Germany for birds that have unusual markings ("colour pigeons"), in Belgium for racing pigeons, and in the United States for dual-purpose breeds. Hundreds of varieties of complicated lineage represent centuries of development.

SQUAB

In culinary terminology, squab is a young domestic pigeon, typically under four weeks old, or its meat. The meat is widely described as tasting like dark chicken. The term is probably of Scandinavian origin; the Swedish word *skvabb* means "loose, fat flesh". It formerly applied to all dove and pigeon species, such as the wood pigeon, the mourning dove, the extinct-in-the-wild socorro dove, and the now-extinct passenger pigeon, and their meat. More recently, squab meat comes almost entirely from domesticated pigeons. The meat of dove and pigeon gamebirds hunted primarily for sport is rarely called squab.

Pigeon chicks, approximately 20 days-of-age.

The practice of domesticating pigeons as livestock may have come from the Middle East; historically, squabs or pigeons have been consumed in many civilizations, including Ancient Egypt (still common in modern Egypt), Rome and Medieval Europe. Although squab has been consumed throughout much of recorded history, it is generally regarded as exotic, not as a contemporary staple food; there are more records of its preparation for the wealthy than for the poor.

The modern squab industry uses utility pigeons. Squabs are raised until they are roughly a month old, when they reach adult size but have not yet flown, before being slaughtered.

A dovecote in the subterranean caves of Orvieto, Italy where since the time of the Etruscans in the Iron Age, the locals raised squab for food.

The practice of domesticating pigeon as livestock may have come from the Middle East; historically, squabs or pigeons have been consumed in many civilizations, including Ancient Egypt, Rome, and Medieval Europe. Doves are described as food in the Holy Scriptures and were eaten by the Hebrews. Texts about methods of raising pigeons for their meat date as far back as AD 60 in Spain. Such birds were hunted for their meat because it was a cheap and readily available source of protein.

In the Tierra de Campos, a resource-poor region of north-western Spain, squab meat was an important supplement to grain crops from at least Roman times. Caelius Aurelianus, an Ancient Roman physician, regarded the meat as a cure for headaches, but by the 16th century, squab was believed to cause headaches.

From the Middle Ages, a dovecote (French *pigeonnier*) was a common outbuilding on an estate that aimed to be self-sufficient. The dovecote was considered a "living pantry", a source of meat for unexpected guests, and was important as a supplementary source of income from the sale of surplus birds. Dovecotes were introduced to South America and Africa by Mediterranean colonists. In medieval England, squab meat was highly valued, although its availability depended on the season—in one dovecote in the 1320s, nearly half the squab yield was produced in the summer, none in the winter.

In England, pigeon meat was eaten when other food was rationed during the Second World War and remains associated with wartime shortages and poverty. This was parodied in an episode of the sitcom *Dad's Army*, "Getting the Bird". Nevertheless, many people continue to eat it, especially the older generation.

A pair of king pigeons. Large breast muscles are common in utility pigeons.

Husbandry

Squab have been commercially raised in North America since the early 1900s. As of 1986, annual production in the United States and Canada was one and a half million squabs per year.

Pigeons, unlike other poultry, form pair bonds to breed, and squabs must be brooded and fed by both parents until they are four weeks old; a pair of pigeons may produce 15 squabs per year. Ten pairs can produce eight squabs each month without being fed by their keepers. Pigeons which are accustomed to their dovecote may forage and return there to rest and breed. Industrially raised pigeons have young which weigh 1.3 pounds (0.59 kg) when of age, as opposed to traditionally raised pigeons, which weigh 0.5 pounds (0.23 kg).

Utility pigeons have been artificially selected for weight gain, quick growth, health when kept in large numbers, and health of their infants. For a greater yield, commercially raised squab may be produced in a two-nest system, where the mother lays two new eggs in a second nest while her offspring are still growing in the first nest, fed crop milk by both parents. Establishing two breeding lines has been suggested as another strategy for greater yield, where one breeding line is selected for prolificacy and the other for "parental performance", which, according to Aggrey and Cheng, is "vital" for squab growth after the age of two weeks.

Meleg estimates that 15–20% of eggs fail to hatch in well-maintained pigeon lofts. Egg size is important for the squab's initial size and for mortality at hatching, but becomes less important as the squab ages. Aggrey and Cheng feel that the hatched weight of squabs is not a good indicator of their weight at four weeks old.

Squabs reach adult size, but are not yet ready to fly (making them easier to catch) after roughly a month; at this point, they are slaughtered.

PEAFOWL

Indian peacock displaying its train.

Peafowl is a common name for three species of birds in the genera *Pavo* and *Afropavo* of the Phasianidae family, the pheasants and their allies. Male peafowl are referred to as peacocks, and female peafowl as peahens. The two Asiatic species are the blue or Indian peafowl originally of the Indian subcontinent, and the green peafowl of Southeast Asia; the one African species is the Congo peafowl, native only to the Congo Basin. Male peafowl are known for their piercing calls and their extravagant plumage. The latter is especially prominent in the Asiatic species, which have an eye-spotted "tail" or "train" of covert feathers, which they display as part of a court-ship ritual.

The functions of the elaborate iridescent colouration and large "train" of peacocks have been the subject of extensive scientific debate. Charles Darwin suggested they served to attract females, and the showy features of the males had evolved by sexual selection. More recently, Amotz Zahavi proposed in his handicap theory that these features acted as honest signals of the males' fitness, since less-fit males would be disadvantaged by the difficulty of surviving with such large and conspicuous structures.

Plumage

Head

The Indian peacock has iridescent blue and green plumage, mostly metallic blue and green, but the green peacock has green and bronze body feathers. In both species, females are as big as males, but lack the train and the head ornament. The peacock "tail", known as a "train", consists not of tail quill feathers, but highly elongated upper tail coverts. These feathers are marked with eyespots, best seen when a peacock fans his tail. Both sexes of all species have a crest atop the head. The Indian peahen has a mixture of dull grey, brown, and green in her plumage. The female also displays her plumage to ward off female competition or signal danger to her young.

Green peafowl differ from Indian peafowl in that the male has green and gold plumage and black wings with a sheen of blue. Unlike Indian peafowl, the green peahen is similar to the male, only having shorter upper tail coverts, a more coppery neck, and overall less iridescence.

The Congo peacock male does not display his covert feathers, but uses his actual tail feathers during courtship displays. These feathers are much shorter than those of the Indian and green species, and the ocelli are much less pronounced. Females of the Indian and African species are dull grey and/or brown.

Chicks of both sexes in all the species are cryptically coloured. They vary between yellow and tawny, usually with patches of darker brown or light tan and "dirty white" ivory.

A leucistic Indian peacock.

Colour and Pattern Variations

Hybrids between Indian and Green peafowl are called Spaldings, after the first person to successfully hybridise them, Mrs. Keith Spalding. Unlike many hybrids, spaldings are fertile and generally benefit from hybrid vigor; spaldings with a high-green pheno-type do much better in cold temperatures than the cold-intolerant green peafowl while still looking like their green parents. Plumage varies between individual spaldings, with some looking far more like green peafowl and some looking far more like blue peafowl, though most visually carry traits of both.

In addition to the wild-type "blue" colouration, several hundred variations in colour and pattern are recognised as separate morphs of the Indian Blue among peafowl breeders. Pattern variations include solid-wing/black shoulder (the black and brown stripes on the wing are instead one solid colour), pied, white-eye (the ocelli in a male's eye feathers have white spots instead of black), and silver pied (a mostly white bird with small patches of colour). Colour variations include white, purple, Buford bronze, opal, midnight, charcoal, jade, and taupe, as well as the sex-linked colours purple, cameo, peach, and Sonja's Violeta. Additional colour and pattern variations are first approved by the United Peafowl Association to become officially recognised as a morph among breeders. Alternately-coloured peafowl are born differently coloured than wild-type peafowl, and though each colour is recognisable at hatch, their peachick plumage does not necessarily match their adult plumage.

Occasionally, peafowl appear with white plumage. Although albino peafowl do exist, this is quite rare, and almost all white peafowl are not albinos; they have a genetic condition called leucism, which causes pigment cells to fail to migrate from the neu-ral crest during development. Leucistic peafowl can produce pigment but not deposit the pigment to their feathers. This results in the complete lack of colouration in their plumage and blue-grey eye colour. Pied peafowl are affected by partial leucism, where only some pigment cells fail to migrate, resulting in birds that have colour but also have patches absent of all colour; they, too, have blue-grey eyes. By contrast, true albino peafowl would have a complete lack of melanin, resulting in irises that look red or pink. Leucistic peachicks are born yellow and become fully white as they mature.

Iridescence

The Royal beauty of the jungle.

As with many birds, vibrant iridescent plumage colours are not primarily pigments, but structural colouration. Optical interference Bragg reflections, based on regular, periodic nanostructures of the barbules (fiber-like components) of the feathers, produce the peacock's colours. Slight changes to the spacing of these barbules result in different colours. Brown feathers are a mixture of red and blue: one colour is created by the periodic structure and the other is created by a Fabry–Pérot interference peak from reflections from the outer and inner boundaries. Such structural colouration causes the iridescence of the peacock's hues. Interference effects depend on light angle rather than actual pigments.

Evolution and Sexual Selection

Charles Darwin suggested in *On the Origin of Species* that the peafowl's plumage had evolved through sexual selection. He expanded upon this in his second book, *The Descent of Man and Selection in Relation to Sex.*

The sexual struggle is of two kinds; in the one it is between individuals of the same sex, generally the males, in order to drive away or kill their rivals, the females remaining passive; whilst in the other, the struggle is likewise between the individuals of the same sex, in order to excite or charm those of the opposite sex, generally the females, which no longer remain passive, but select the more agreeable partners.

Sexual selection is the ability of male and female organisms to exert selective forces on each other with regard to mating activity. The strongest driver of sexual selection is gamete size. In general, eggs are bigger than sperm, and females produce fewer gametes than males. This leads to eggs being a bigger investment, so to females being choosy about the traits that will be passed on to her offspring by males. The peahen's reproductive success and the likelihood of survival of her chicks is partly dependent on the genotype of the mate. Females generally have more to lose when mating with an inferior male due to her gametes being more costly than the male's.

Female Choice

Peacock (seen from behind) displaying to attract peahen in foreground.

Multiple hypotheses attempt to explain the evolution of female choice. Some of these suggest direct benefits to females, such as protection, shelter, or nuptial gifts that sway the female's choice of mate. Another hypothesis is that females choose mates with good genes. Males with more exaggerated secondary sexual characteristics, such as bigger, brighter peacock trains, tend to have better genes in the peahen's eyes. These better genes directly benefit her offspring, as well as her fitness and reproductive success. Runaway selection also seeks to clarify the evolution of the peacock's train. In runaway sexual selection, linked genes in males and females code for sexually dimorphic traits in males, and preference for those traits in females. The close spatial association of alleles for loci involved in the train in males, and for preference for more exuberant trains in females, on the chromosome (linkage disequilibrium) causes a positive feedback loop that exaggerates both the male traits and the female preferences. Another hypothesis is sensory bias, in which females have a preference for a trait in a nonmating context that becomes transferred to mating. Multiple causality for the evolution of female choice is also possible.

Work concerning female behavior in many species of animals has sought to confirm Darwin's basic idea of female preference for males with certain characteristics as a major force in the evolution of species. Females have often been shown to distinguish small differences between potential mates, and to prefer mating with individuals bearing the most exaggerated characters. In some cases, those males have been shown to be more healthy and vigorous, suggesting that the ornaments serve as markers indicating the males' abilities to survive, and thus their genetic qualities.

The peacock's train and iridescent plumage are perhaps the best-known example of traits believed to have arisen through sexual selection, though with some controversy. Male peafowl erect their trains to form a shimmering fan in their display to females. Marion Petrie tested whether or not these displays signalled a male's genetic quality by studying a feral population of peafowl in Whipsnade Wildlife Park in southern England. The number of eyespots in the train predicted a male's mating success. She was able to manipulate this success by cutting the eyespots off some of the males' tails: females lost interest in pruned males and became attracted to untrimmed ones. Males with fewer eyespots, thus with lower mating success, suffered from greater predation. She allowed females to mate with males with differing numbers of eyespots, and reared the offspring in a communal incubator to control for differences in maternal care. Chicks fathered by more ornamented males weighed more than those fathered by less ornamented males, an attribute generally associated with better survival rate in birds. These chicks were released into the park and recaptured one year later. Those with heavily ornamented feathers were better able to avoid predators and survive in natural conditions. Thus, Petrie's work has shown correlations between tail ornamentation, mating success, and increased survival ability in both the ornamented males and their offspring.

A peacock in flight: Zahavi argued that the long train would be a handicap.

Furthermore, peafowl and their sexual characteristics have been used in the discussion of the causes for sexual traits. Amotz Zahavi used the excessive tail plumes of male peafowls as evidence for his "handicap principle". Since these trains are likely to be deleterious to the survival of an individual (as the brilliant plumes are visible to predators and the longer plumes make escape from danger more difficult), Zahavi argued that only the fittest males could survive the handicap of a large train. Thus, a brilliant train serves as an honest indicator for females that these highly ornamented males are good at surviving for other reasons, so are preferable mates. This theory may be contrasted with Ronald Fisher's theory (and Darwin's hypothesis) that male sexual traits are the result of initially arbitrary aesthetic selection by females.

In contrast to Petrie's findings, a seven-year Japanese study of free-ranging peafowl concluded that female peafowl do not select mates solely on the basis of their trains. Mariko Takahashi found no evidence that peahens preferred peacocks with more elaborate trains (such as with more eyespots), a more symmetrical arrangement, or a greater length. Takahashi determined that the peacock's train was not the universal target of female mate choice, showed little variance across male populations, and did not cor-relate with male physiological condition. Adeline Loyau and her colleagues responded that alternative and possibly central explanations for these results had been overlooked. They concluded that female choice might indeed vary in different ecological conditions.

Food Courtship Theory

Merle Jacobs' food-courtship theory states that peahens are attracted to peacocks for the resemblance of their eye spots to blue berries.

Natural Selection

It has been suggested that a peacock's train, loud call, and fearless behavior have been formed by natural selection (not sexual selection), and served as an aposematic display to intimidate predators and rivals.

Plumage Colours as Attractants

Eyespot on a peacock's train feather.

A peacock's copulation success rate depends on the colours of his eyespots (ocelli) and the angle at which they are displayed. The angle at which the ocelli are displayed during courtship is more important in a peahen's choice of males than train size or number of ocelli. Peahens pay careful attention to the different parts of a peacock's train during his display. The lower train is usually evaluated during close-up courtship, while the upper train is more of a long-distance attraction signal. Actions such as train rattling and wing shaking also kept the peahens' attention.

Redundant Signal Hypothesis

Although an intricate display catches a peahen's attention, the redundant signal hy-pothesis also plays a crucial role in keeping this attention on the peacock's display.

The redundant signal hypothesis explains that whilst each signal that a male projects is about the same quality, the addition of multiple signals enhances the reliability of that mate. This idea also suggests that the success of multiple signalling is not only due to the repetitiveness of the signal, but also of multiple receivers of the signal. In the peacock species, males congregate a communal display during breeding season and the peahens observe. Peacocks first defend their territory through intra-sexual behavior, defending their areas from intruders. They fight for areas within the congregation to display a strong front for the peahens. Central positions are usually taken by older, dominant males, which influences mating success. Certain morphological and behavioral traits come in to play during inter and intra-sexual selection, which include train length for territory acquisition and visual and vocal displays involved in mate choice by peahens.

Vocalisation

In courtship, vocalisation stands to be a primary way for peacocks to attract peahens. Some studies suggest that the intricacy of the "song" produced by displaying peacocks proved to be impressive to peafowl. Singing in peacocks usually occurs just before, just after, or sometimes during copulation.

Behavior

Peacock sitting.

Peafowl are forest birds that nest on the ground, but roost in trees. They are terrestrial feeders. All species of peafowl are believed to be polygamous. In common with other members of the Galliformes, the males possess metatarsal spurs or "thorns" on their legs used during intraspecific territorial fights with other members of their kind.

A green peafowl (*Pavo muticus*).

Diet

Peafowl are omnivores and eat mostly plant parts, flower petals, seed heads, insects and other arthropods, reptiles, and amphibians. Wild peafowl look for their food scratching around in leaf litter either early in the morning or at dusk. They retreat to the shade and security of the woods for the hottest portion of the day. These birds are not picky and will eat almost anything they can fit in their beak and digest. They actively hunt insects like ants, crickets and termites; millipedes; and other arthropods and small mammals. Indian peafowl also eat small snakes.

Domesticated peafowl may also eat bread and cracked grain such as oats and corn, cheese, cooked rice and sometimes cat food. It has been noticed by keepers that peafowl enjoy protein rich food including larvae that infest granaries, different kinds of meat and fruit, as well as vegetables including dark leafy greens, broccoli, carrots, beans, beets, and peas.

DOMESTIC TURKEYS

A broad breasted bronze male (tom) displaying.

The domestic turkey (*Meleagris gallopavo*) is a large fowl, one of the two species in the genus *Meleagris* and the same as the wild turkey. Although turkey domestication was thought to have occurred in central Mesoamerica at least 2,000 years ago, recent research suggests a possible second domestication event in the Southwestern United States between 200 BC and AD 500. However, all of the main domestic turkey varieties today descend from the turkey raised in central Mexico that was subsequently imported into Europe by the Spanish in the 16th century.

Domestic turkey is a popular form of poultry, and it is raised throughout temperate parts of the world, partially because industrialized farming has made it very cheap for the amount of meat it produces. Female domestic turkeys are referred to as *hens*, and the chicks may be called *poults* or *turkeylings*. In the United States, the males are referred to as *toms*, while in the United Kingdom and Ireland, males are *stags*.

The great majority of domestic turkeys are bred to have white feathers because their pin feathers are less visible when the carcass is dressed, although brown or bronze-feathered varieties are also raised. The fleshy protuberance atop the beak is the snood, and the one attached to the underside of the beak is known as a wattle.

The English language name for this species results from an early misidentification of the bird with an unrelated species which was imported to Europe through the country of Turkey.

Behavior

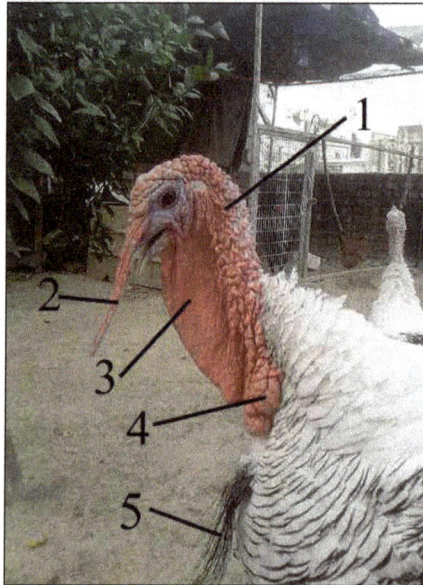

Anatomical structures on the head and throat of a domestic turkey. 1. Caruncles,
2. Snood, 3. Wattle (Dewlap), 4. Major caruncle, 5. Beard.

Young domestic turkeys readily fly short distances, perch and roost. These behaviors become less frequent as the birds mature, but adults will readily climb on objects such as bales of straw. Young birds perform spontaneous, frivolous running ('frolicking') which has all the appearance of play. Commercial turkeys show a wide diversity of behaviors including 'comfort' behaviors such as wing-flapping, feather ruffling, leg stretching and dust-bathing. Turkeys are highly social and become very distressed when isolated. Many of their behaviors are socially facilitated i.e. expression of a behavior by one animal increases the tendency for this behavior to be performed by others. Adults can recognise 'strangers' and placing any alien turkey into an established group will almost certainly result in that individual being attacked, sometimes fatally. Turkeys are highly vocal, and 'social tension' within the group can be monitored by the birds' vocalisations. A high-pitched trill indicates the birds are becoming aggressive which can develop into intense sparring where opponents leap at each other with the large, sharp talons, and try to peck or grasp the head of each other. Aggression increases in frequency and severity as the birds mature.

Male domestic turkey sexually displaying by showing the snood hanging over the beak, the caruncles hanging from the throat, and the 'beard' of small, black, stiff feathers on the chest.

Maturing males spend a considerable proportion of their time sexually displaying. This is very similar to that of the wild turkey and involves fanning the tail feathers, drooping the wings and erecting all body feathers, including the 'beard' (a tuft of black, modified hair-like feathers on the centre of the breast). The skin of the head, neck and caruncles (fleshy nodules) becomes bright blue and red, and the snood (an erectile appendage on the forehead) elongates, the birds 'sneeze' at regular intervals, followed by a rapid vibration of their tail feathers. Throughout, the birds strut slowly about, with the neck arched backward, their breasts thrust forward and emitting their characteristic 'gobbling' call.

Size and Weight

Animal	Average mass kg (lb)	Maximum mass kg (lb)	Average total length cm (ft)
Domestic turkey	13.5 (29.8)	39 (86)	100 - 124.9 (3.3 – 4.1)

The domestic turkey is the eighth largest living bird species in terms of maximum mass at 39 kg (86 lbs).

Turkey Breeds

A domestic turkey taken as a pet.

- The Broad Breasted White is the commercial turkey of choice for large scale industrial turkey farms, and consequently is the most consumed variety of the bird. Usually the turkey to receive a "presidential pardon", a U.S. custom, is a Broad Breasted White.

- The Broad Breasted Bronze is another commercially developed strain of table bird.

- The Standard Bronze looks much like the Broad Breasted Bronze, except that it is single breasted, and can naturally breed.

- The Bourbon Red turkey is a smaller, non-commercial breed with dark reddish feathers with white markings.

- Slate, or Blue Slate, turkeys are a very rare breed with gray-blue feathers.

- The Black ("Spanish Black", "Norfolk Black") has very dark plumage with a green sheen.

- The Narragansett Turkey is a popular heritage breed named after Narraganset Bay in New England.

- The Chocolate is a rarer heritage breed with markings similar to a Black Spanish, but light brown instead of black in color. Common in the Southern U.S. and France before the Civil War.

- The Beltsville Small White is a small heritage breed, whose development started in 1934. The breed was introduced in 1941 and was admitted to the APA Standard in 1951. Although slightly bigger and broader than the Midget White, both are often mislabeled.

- The Midget White is a smaller heritage breed.

Commercial Production

Turkey in Pakistan.

In commercial production, breeder farms supply eggs to hatcheries. After 28 days of incubation, the hatched poults are sexed and delivered to the grow-out farms; hens are raised separately from toms because of different growth rates.

In the UK, it is common to rear chicks in the following way. Between one and seven days of age, chicks are placed into small 2.5 m (8 ft) circular brooding pens to ensure they encounter food and water. To encourage feeding, they may be kept under constant light for the first 48 hours. To assist thermoregulation, air temperature is maintained at 35 °C (95 °F) for the first three days, then lowered by approximately 3 °C (5.4 °F) every two days to 18 °C (64 °F) at 37 days of age, and infra-red heaters are usually provided for the first few days. Whilst in the pens, feed is made widely accessible by scattering it on sheets of paper in addition to being available in feeders. After several days, the pens are removed, allowing the birds access to the entire rearing shed, which may contain tens of thousands of birds. The birds remain there for several weeks, after which they are transported to another unit.

The vast majority of turkeys are reared indoors in purpose-built or modified buildings of which there are many types. Some types have slatted walls to allow ventilation, but many have solid walls and no windows to allow artificial lighting manipulations to optimise production. The buildings can be very large (converted aircraft hangars are sometimes used) and may contain tens of thousands of birds as a single flock. The floor substrate is usually deep-litter, e.g. wood shavings, which relies upon the controlled build-up of a microbial flora requiring skilful management. Ambient temperatures for adult domestic turkeys are usually maintained between 18 and 21 °C (64 and 70 °F). High temperatures should be avoided because the high metabolic rate of turkeys (up to 69 W/bird) makes them susceptible to heat stress, exacerbated by high stocking densities. Commercial turkeys are kept under a variety of lighting schedules, e.g. continuous light, long photoperiods (23 h), or intermittent lighting, to encourage feeding and accelerate growth. Light intensity is usually low (e.g. less than one lux) to reduce feather pecking.

Rations generally include corn and soybean meal, with added vitamins and minerals, and is adjusted for protein, carbohydrate and fat based on the age and nutrient requirements. Hens are slaughtered at about 14–16 weeks and toms at about 18–20 weeks of age when they can weigh over 20 kg (44 lb) compared to a mature male wild turkey which weighs approximately 10.8 kg (24 lb).

Welfare Concerns

Modern domestic turkeys under commercial conditions.

Stocking density is an issue in the welfare of commercial turkeys and high densities are a major animal welfare concern. Permitted stocking densities for turkeys reared indoors vary according to geography and animal welfare farm assurance schemes. For example, in Germany, there is a voluntary maximum of 52 kg/m² and 58 kg/m² for males and females respectively. In the UK, the RSPCA Freedom Foods assurance scheme reduces permissible stocking density to 25 kg/m² for turkeys reared indoors. Turkeys maintained at commercial stocking densities (8 birds/m²; 61 kg/m²) exhibit increased welfare problems such as increases in gait abnormalities, hip and foot lesions, and bird disturbances, and decreased bodyweight compared with lower stocking densities. Turkeys reared at 8 birds/m² have a higher incidence of hip lesions and foot pad dermatitis than those reared at 6.5 or 5.0 birds/m². Insufficient space may lead to an increased risk for injuries such as broken wings caused by hitting the pen walls or other turkeys during aggressive encounters and can also lead to heat stress. The problems of small space allowance are exacerbated by the major influence of social facilitation on the behavior of turkeys. If turkeys are to feed, drink, dust-bathe, etc., simultaneously, then to avoid causing frustration, resources and space must be available in large quantities.

Lighting manipulations used to optimise production can compromise welfare. Long photoperiods combined with low light intensity can result in blindness from buphthalmia (distortions of the eye morphology) or retinal detachment.

Feather pecking occurs frequently amongst commercially reared turkeys and can begin at 1 day of age. This behavior is considered to be re-directed foraging behavior, caused by providing poultry with an impoverished foraging environment. To reduce feather pecking, turkeys are often beak-trimmed. Ultraviolet-reflective markings appear on young birds at the same time as feather pecking becomes targeted toward these areas, indicating a possible link. Commercially reared turkeys also perform head-pecking, which becomes more frequent as they sexually mature. When this occurs in small enclosures or environments with few opportunities to escape, the outcome is often fatal and rapid. Frequent monitoring is therefore essential, particularly of males approaching maturity. Injuries to the head receive considerable attention from other birds, and head-pecking often occurs after a relatively minor injury has been received during a fight or when a lying bird has been trodden upon and scratched by another. Individuals being re-introduced after separation are often immediately attacked again. Fatal head-pecking can occur even in small (10 birds), stable groups. Commercial turkeys are normally reared in single-sex flocks. If a male is inadvertently placed in a female flock, he may be aggressively victimised (hence the term 'henpecked'). Females in male groups will be repeatedly mated, during which it is highly likely she will be injured from being trampled upon.

Breeding and Companies

The dominant commercial breed is the Broad-breasted Whites (similar to "White Holland", but a separate breed), which have been selected for size and amount of meat. Mature toms are too large to achieve natural fertilization without injuring the hens, so their semen is collected, and hens are inseminated artificially. Several hens can be inseminated from each collection, so fewer toms are needed. The eggs of some turkey breeds are able to develop without fertilization, in a process called parthenogenesis. Breeders' meat is too tough for roasting, and is mostly used to make processed meats.

Waste Products

Approximately two billion to four billion pounds (900,000 to 1,800,000 t) of poultry feathers are produced every year by the poultry industry. Most are ground into a protein source for ruminant animal feed, which are able to digest the protein keratin of which feathers are composed. Researchers at the United States Department of Agriculture (USDA) have patented a method of removing the stiff quill from the fibers which make up the feather. As this is a potential supply of natural fibers, research has been conducted at Philadelphia University's School of Engineering and Textiles to determine textile applications for feather fibers. Turkey feather fibers have been blended with nylon and spun into yarn, and then used for knitting. The yarns were tested for strength while the fabrics were evaluated as insulation materials. In the case of the yarns, as the percentage of turkey feather fibers increased, the strength decreased. In fabric form, as the percentage of turkey feather fibers increased, the heat retention capability of the fabric increased.

Turkey Litter for Fuel

Although most commonly used as fertilizer, turkey litter (droppings mixed with bedding material, usually wood chips) is used as a fuel source in electric power plants. One such plant in western Minnesota provides 55 megawatts of power using 500,000 tons of litter per year. The plant began operating in 2007.

GUINEA FOWL

Guinea fowl is part of a family, Numididae (order Galliformes), of African birds that are alternatively placed by some authorities in the pheasant family, Phasianidae. The family consists of 7–10 species, one of which, Numida meleagris, is widely domesticated for its flesh and as a "watchdog" on farms (it gabbles loudly at the least alarm). The largest and most-colourful species is the vulturine guinea fowl (Acryllium vulturinum), of eastern Africa, a long-necked bird with a hackle of long lance-shaped feathers striped black, white, and blue; red eyes; and a vulturelike bare blue head.

Wild forms of N. meleagris are known as helmet guinea fowl from their large bony crest; the sexes look alike. The helmet guinea fowl has many local varieties widespread in the savannas and scrublands of Africa and has been introduced into the West Indies and elsewhere. About 50 cm (20 inches) long, the typical form has a bare face, brown eyes, red and blue wattles at the bill, black white-spotted plumage, and hunched posture. It lives in flocks and walks about on the ground, feeding on seeds, tubers, and some insects. When alarmed the birds run, but when pressed they fly on short, rounded wings for a short distance. At night they sleep in trees. Helmet guinea fowl are noisy birds giving harsh, repetitive calls. The nest is a hollow in the ground and is scantily lined with vegetation. It contains about 12 finely spotted tan-coloured eggs, which require about 30 days' incubation. The downy young are active immediately after hatching and accompany the parents.

Helmet guinea fowl (Numida meleagris).

References

- Buffetaut, E.; Angst, D. (November 2014). "Stratigraphic distribution of large flightless birds in the Palaeogene of Europe and its palaeobiological and palaeogeographical implications". Earth-Science Reviews. 138: 394–408. Doi:10.1016/j.earscirev.2014.07.001

- Chicken, entry: newworldencyclopedia.org, Retrieved 27 March, 2019

- Mikko's Phylogeny Archive [1] Haaramo, Mikko (2007). "Paleognathia - paleognathous modern birds". Retrieved 30 December 2015

- Guinea-fowl, animal: britannica.com, Retrieved 28 April, 2019

- "Squab". Merriam-Webster's Collegiate Dictionary (11th ed.). 2004. P. 1210. ISBN 978-0-87779-809-5. Retrieved 27 August 2009

- Blau, S. K. (January 2004). "Light as a Feather: Structural Elements Give Peacock Plumes Their Colour". Physics Today. 57 (1): 18–20. Bibcode:2004pht....57a..18B. Doi:10.1063/1.1650059. Archived from the original on 22 June 2006

3

Poultry Farming

The form of animal husbandry which is involved in raising domesticated birds such as chickens, ducks, turkeys and geese to produce meat or eggs for food is termed as poultry farming. Some of the birds which are raised under poultry farming are ducks, broiler chickens and peacocks. The chapter closely examines the farming of these types of poultry to provide an extensive understanding of the subject.

Poultry farming is the raising of birds domestically or commercially, primarily for meat and eggs but also for feathers. Chickens, turkeys, ducks, and geese are of primary importance, while guinea fowl and squabs (young pigeons) are chiefly of local interest.

Commercial Production

Feeding

Commercial poultry feeding is a highly perfected science that ensures a maximum intake of energy for growth and fat production. High-quality and well-balanced protein sources produce a maximum amount of muscle, organ, skin, and feather growth. The essential minerals produce bones and eggs, with about 3 to 4 percent of the live bird being composed of minerals and 10 percent of the egg. Calcium, phosphorus, sodium, chlorine, potassium, sulfur, manganese, iron, copper, cobalt, magnesium, and zinc are all required. Vitamins A, C, D, E, and K and all of the B vitamins are also required. Antibiotics are widely used to stimulate appetite, control harmful bacteria, and prevent disease. For chickens, modern rations produce about 0.5 kg (1 pound) of broiler on about 0.9 kg (2 pounds) of feed and a dozen eggs from 2 kg (4.5 pounds) of feed.

Young chickens in an organic poultry farm.

Management

A carefully controlled environment that avoids crowding, chilling, overheating, or frightening is almost universal in poultry farming. Cannibalism, which expresses itself as toe picking, feather picking, and tail picking, is controlled by debeaking at one day of age and by other management practices. The feeding, watering, egg gathering, and cleaning operations are highly mechanized. Birds are usually housed in wire cages with two or three animals per cage, depending on the species and breed, and three or four tiers of cages superposed to save space. Cages for egg-laying birds have been found to increase production, lower mortality, reduce cannibalism, lower feeding requirements, reduce diseases and parasites, improve culling, and reduce both space and labour requirements.

Poultry breeding is an outstanding example of the application of basic genetic principles of inbreeding and crossbreeding as well as of intensive mass selection to effect faster and cheaper gains in meat and maximum egg production for the egg-laying strains. Maximum use of heterosis, or hybrid vigour, through incrosses and crossbreeding has been made. Rapid and efficient weight gains and high-quality, plump, meaty carcasses have been achieved thereby. Among the world's agricultural industries, chicken breeding in the U.S. is one of the most advanced. Intensive nutritional research and application, highly improved breeding stock, intelligent management, and scientific disease control have gone into the effort to give a modern broiler (meat chicken) of uniformly high quality produced at ever-lower cost. A modern broiler chick can reach a 2.3-kg (5-pound) market weight in five weeks, compared with the four months that were required in the mid-20th century. Additionally, annual egg production per hen has increased from about 100 in 1910 to over 300 in the early 21st century.

Diseases

Poultry are quite susceptible to a number of diseases; some of the more common are fowl typhoid, pullorum, fowl cholera, chronic respiratory disease, infectious sinusitis, infectious coryza, avian infectious hepatitis, infectious synovitis, bluecomb, Newcastle disease, fowl pox, avian leukosis complex, coccidiosis, blackhead, infectious laryngotracheitis, infectious bronchitis, and erysipelas. Strict sanitary precautions, the intelligent use of antibiotics and vaccines, and the widespread use of cages for layers and confinement rearing for broilers have made it possible to effect satisfactory disease control.

Parasitic diseases of poultry, including hexamitiasis of turkeys, are caused by round-worms, tapeworms, lice, and mites. Again, modern methods of sanitation, prevention, and treatment provide excellent control.

DUCK FARMING

Duck farming is very popular and absolutely a lucrative business. Ducks are highly available around the world. There are numerous meat and egg productive duck breeds available throughout the world. All the present domestic ducks around the glove come from the wild birds.

Those wild birds wonder around the world, and some of them have been domesticated as a good source of food. Almost all those wild birds are from mallard species. Incidentally, all hen of the world comes from red wild hen. You might know that, ducks are aquatic organism.

Some people think that, duck without water and pond without water are the same. Even, some people think, ducks can't live without water. But it's totally wrong. You can not imagine pond without water, but duck can be raised without water. Thousands of ducks can be raised without water by keeping them inside house, in the same way you raise chickens or other types of poultry birds.

But, keep in mind that, in case of raising ducks without water 'your ducks will lay un-fertilized egg'. That means you can't hatch the eggs for producing ducklings. If you want fertile eggs, male ducks and water are essential. You can easily raise ducks without wa-ter, they just need water for reproduction or mating purposes.

Advantages of Duck Farming

There are numerous advantages of starting duck farming business. In many countries, ducks rank next to chicken for meat and egg production. You can raise ducks in both commercial and small scale meat or egg production purpose. Even, you can raise some ducks on your own backyard with other birds or animals. Some notable advantages of duck farming business are:

- Ducks need less expensive, simple and non-elaborate housing facilities. As a re-sult housing costs are very less for setting up commercial duck farming business.

- Ducks are very hardy bird and they need less care or management. They can adopt themselves with almost all types of environmental conditions.

- They lay eggs either at night or in the morning. So you can collect their fresh eggs every morning . And you can do your other work during rest of the day and you don't have to spend time for caring your ducks.

- You need comparatively less space for raising ducks. Ducks have comparatively shorter brooding period and ducklings grow faster. Ducklings grow so fast that, you can dispense artificial heat within their 5 to 7 days. Although they will require a little longer heating period during cold months.

- Ducks are highly resistant to the common avian diseases.

- You can feed your ducks with a wide variety of foods. A duck's regular food includes cassava, copra, corn, rice, fruits and any other low cost and easily available foods. They also have the natural tendency of foraging on aquatic weeds, algae, green legumes, fungi, earthworms, maggots, snails, various types of insects etc. which directly reduce feeding cost.

- You can also use your ducks for controlling apple snails or some other harmful insects from your garden.

- Ducks have less mortality rate and usually they live longer than chickens. In case of egg production, ducks lay eggs for a long time period.

- Duck products such as eggs and meat have a great demand in the local and international market. So commercial duck farming business can be a great source of earning. There are already many successful farmers who are making a high profit from their duck farming business.

- Duck farming business can also be a stable employment source. Young unemployed educated people can join this business and make their own employment source.

Peculiarity of Ducks

The body of the ducks is fully covered with oily feathers, so water can't directly enter inside their body. Ducks have a layer of fat under their skin, which helps them from getting wet. For this reason, if ducks spend a couple of hours in water, their body remain dry.

Another peculiarity of duck is their legs. There are three fingers on their legs and covered with thin skin (which is called 'web'). It works as the paddle of a boat. Finger ends in nail. Some species of ducks use their nails for defending themselves. Generally the lip of ducks are red or orange colored and become very strong. With this lip they can eat moss, insects, fish egg, hard snail etc. They send almost everything inside their stomach, which they found between their lip.

Usually hens lay egg all over the day, but ducks lay eggs at night or in the morning. So don't allow your ducks to go out from their house before 9 am. If you observe ducks very closely, you will find that they are very clever and intelligent. You can spent your leisure time easily by watching their activities.

Duck's Feed

Ducks generally eat almost all types of food they find edible. You can feed your ducks like chickens. But you have to add some extra additives into duck feed. As some ducks lay more eggs than hens, so you have to be very careful about feeding your ducks. Add necessary nutrient elements in their diet.

Always provide your ducks nutritious feed according to the various types of duck breeds and their growth rate. For small scale or domestic duck farming, you can feed your ducks rich bran, kitchen waste and a plenty of snails. Always keep in mind that, ducks eat a lot of food daily. So you have to feed them well-balanced food if you want proper egg and meat production.

Duck's Eggs and Meat Market

Marketing duck products (egg and meat) is very easy. People all over the world like duck egg and meat from the ancient time. In some areas, people like duck eggs and meat more than chicken meat or eggs. Duck eggs are usually larger in size than chicken eggs. In some areas duck meat is less popular than their eggs. However, there is no tension about the marketing of ducks eggs and meat. You can try to determine the demand of duck products in your local market.

Duck Breeds

There are numerous duck breeds available throughout the world. Although all of those breeds are not suitable for commercial duck farming business. Some of those breeds are suitable for egg production and some are ideal for commercial meat production. Usually ducks are of three types according to their production type.

- Meat productive duck breeds.

- Egg productive duck breeds.

- Famous for both meat and egg production.

Meat Duck Breeds

There are numerous duck breeds available which are famous for meat production. Peking, Ayleshbari, Maskovi, Ruel Kagua and the Swiden ducks are most popular for meat production. Usually meat productive male ducks weight about 5 kg and female weight about 4 kg.

Housing

The main advantages of duck farming is their simple accommodation. Making a duck house is very easy. You can keep your ducks in low, high, wet, dry and any other places. Ducks like watery and wet place to live. You can easily make a suitable place for your ducks by using a big fruit basket, wood-speech or oil drum. Just put these in a suitable place. Always make a door for entrancing and exiting, wherever you keep the ducks. The door need to be high enough. Because duck enter inside the house raising their head higher. Their house must have to be well ventilated and you have to ensure sufficient flow of fresh air inside the house. Usually each duck will require 2 to 3 square feet flooring space. You also have to be ensure the safety of your ducks. Keep them free from all types of harmful animals or predators, especially from fox and dog. If you want to keep your ducks like chickens, in a concrete floored house, then you have to make a deep liter inside the house. And it will be better if the liter depth between 5-6 inches. In this system, your ducks will lay eggs on the floor.

Breeding

Water is a must for breeding purpose. Usually ducks don't mate without water. One male duck is sufficient for breeding 10 female ducks. Usually high quality and productive duck breeds start laying eggs at their five months of age. Each egg weights about 50 to 60 grams. You can use hens for hatching the eggs, instead of ducks. Do not keep the newly born ducklings in water during their first week, after birth. Because they can get cold if you allow them to water. Usually duck eggs take about 28 days to hatch. During the hatching period, sprinkle the eggs with water occasionally (two or three times per week). You can use both electronic or diesel incubators for hatching their eggs. But many farmer prefer hens than incubator for hatching.

Storing Eggs

You have to store the eggs properly. You can use refrigerator for storing the eggs. Lime water or pyraphine can be used for egg storing purpose. Some people also use actinium glass water.

Gender Determination

It is not so difficult to determine the gender of newly born duckling. Raise up the tail of the duckling and press on its ass. If you notice there is a penis like thorn, then it will be male. And if not then it will definitely be a female duck. The easiest way to determine male ducks is, the tail feathers of male ducks are curled up.

Marketing

Before marketing duck eggs, clean the dirt of egg shell perfectly. Don't wash the eggs with water. You can clean the eggs with knife, soiled paper or towel. Use egg basket or carton for caring eggs from one place to another place. You should choose such cartoon so that each cartoon can contain at least 30 eggs. You can also use bamboo basket, wooden box and other things for transporting eggs to the market or another destination. Make a deep layer of straw or rice bran if you use such boxes for caring eggs.

However, for making maximum profits from duck farming business, you have to be more careful on taking care of ducks, feed management, accommodation, brooding and marketing. If all the process are done well, then you can make a handsome income from this business.

BROILER INDUSTRY

The broiler industry is the process by which broiler chickens are reared and prepared for meat consumption.

A broiler chicken in Ecuador.

Broiler Industry Structure

The broiler production process is very much an industrial one. There are several distinct components of the broiler supply chain.

Primary Breeding Sector

The "primary breeding sector" consists of companies that breed pedigree stock. Pedigree stock ("pure line") is kept on high level biosecure farms. Eggs are hatched in a special pedigree hatchery and their progeny then goes on to the great grandparent (GGP) and grandparent (GP) generations. These eggs would then go to a special GP hatchery to produce Parent Stock (PS) which passes to the production sector.

In 2006, out of an estimated world population of 18 billion poultry, about 3% are breeding stock. The US supplied about 1/4 of world GP stock.

Worldwide, the primary sector produced 417 million parent stock (PS) per year.

A single pedigree-level hen might have 25000 parent stock bird descendants, which in turn might produce 3 million broilers.

Numerous techniques are used to assess the pedigree stock. For example, birds might be examined with ultrasound or x-rays to study the shape of muscles and bones. The blood oxygen level is measured to determine cardiovascular health. The walking ability of pedigree candidates is observed and scored.

The need for high levels of R&D spending prompted consolidation within the primary breeder industry. As of 2017, only two sizable breeding groups remained:

- Aviagen (with the Ross, Hubbard, Arbor Acres, Indian River and Peterson brands).

- Cobb-Vantress (with the Cobb, Avian, Sasso and Hybro brands).

In the UK, 2 international firms supply about 90% of the parent stock.

Due to the high levels of variation in the chicken genome, the industry has not yet reached biological limits to improved performance.

The full chicken genome was published in Nature, in December 2004. Today, all primary breeding groups are investing heavily in genomics research. This research mostly focuses on understanding the function and effect of genes already present in the breeding population. Research into transgenics — removing genes or artificially moving genes from one individual or species to another — has fewer prospects of gaining favor among consumers.

Broiler Breeder Farms

Broiler breeder farms raise parent stock which produce fertilized eggs. A broiler hatching egg is never sold at stores and is not meant for human consumption. The males and females are separate genetic lines or breeds. The chicks they produce will therefore be crossbreeds or "crosses". Since the birds are bred mainly for efficient meat production, producing eggs can be a challenge. In Canada, the average producer houses 15,000 birds that begin laying hatching eggs at 26 weeks of age. Each bird will lay about 150 hatching eggs for the next 34 to 36 weeks. This cycle is then repeated when the producer puts another flock of 26 week-old birds into his barns to begin the process again. As a general rule, each farmer produces enough broiler hatching eggs to supply chicks for 8 chicken producers. (Other sources indicate a parent hen will lay about 180 eggs in a 40-week production period).

Generally, parent flocks are either owned by integrated broiler companies or hatcheries or are contracted to them on a long-term basis.

Broiler breeder growing is typically a two-stage process. Parent stock purchased from a primary breeder is delivered as day old. Most are first placed with on specialist *rearing houses* or starter farms until approximately 18 weeks of age. The starter farm has the specialized brooding equipment to raise the chicks.

Rearing House

Florida chicken house.

A typical rearing house (also called a shed or barn) design for Alabama-like climate (100 °F (38 °C) in summer and 20 °F (−7 °C) in winter):

- 40 by 400 feet (12 m × 122 m) size, single storey.

- 11,000 bird capacity (about 1.4 sq ft (0.13 m²) per bird).

- Ceiling is insulated.

- Exterior curtain side walls.

- A "minimum ventilation" system is required for the heating period to provide a certain amount of fresh air.

- A separate "tunnel ventilation" system with evaporative pad cooling is desired (minimum wind speed is 400 fpm) for hot weather in the later stage of the bird's growth.

- Air inlets may be automatically adjusted.

- A negative ventilation system helps keep dirt and dust out of egg storage areas.

- The entire house may be heated, or individual "brooders" may be used.

- The floor is flat. There are no "slats" or "pits" for manure. There are no cages, and no nests. "Litter" (shavings or straw) covers the floor. When the chicks are introduced temporary barriers are used to keep them close to the heated areas.

- "Black-out" design to keep out external light, so the day-night cycle can be controlled.

- An automatic timer-controlled lighting system. Dimmers allow light intensity to be adjusted.

- Automatic feeders to distribute feed. Typically this consists of an endless chain in a trough or with individual pans. A silo or bin outside provides storage.

- Automatic drinkers provide water. There are several different designs, with "nipples" or "round" drinkers being popular.

- Feeders and drinkers are height adjusted as the birds grow, and can be raised on chains or wires to allow cleanout of the barn.

Chicks require warm air temperatures, which is reduced as the birds mature:

Age	Brooder Temperature	Whole-House Heating Temperature
0 days	34–35 °C (93-95 °F)	31–32 °C (88-90 °F)
14 days	31–32 °C (88-90 °F)	24–25 °C (75-77 °F)

Chicks might be debeaked at 7–10 days age. During rearing, bird weight is carefully monitored, as an over-weight bird will be a poor egg producer. The feed mix will be adjusted to meet nutritional needs at each stage. Feed might be restricted to control body weight, for example with "skip a day" feeding, or feeding 5 days out of 7. A vaccination program is carried out, which ensures the longevity of the parent stock, and the immunity may be passed to the broiler progeny. Males (cockerels) and females (pullets), are usually raised separately.

Laying House

The birds are then moved to broiler breeder *laying houses* or production barns. The birds are typically placed into crates, and transported by truck to a separate facility. Males and females are raised together at this point. Outwardly the laying house will resemble the rearing house. Inside, about one-half of the floor might consist of raised 'slats.' During the production run, manure will drop through the slats and accumulate in the pit underneath the slats. The birds are not generally caged, especially since the roosters must mate with the hens to fertilize the eggs. Nests are provided for laying hens. Both automatic and manual nesting systems exist. Manual nests are usually stuffed with straw or shavings and eggs are hand-collected. Automatic systems usually have a plastic carpet lining, with a belt for egg collection. Careful layout and attention to bird behavior is required to avoid 'floor eggs'.

Depending on breed, egg production starts at 24–26 weeks of age. Production percentage (daily eggs per hen) climbs rapidly to a peak of 80–85% at 29–32 weeks, and then gradually declines with age. Hatchability tends to peak (at perhaps 90%) somewhat later than production at 34–36 weeks. Overall flock production will decline as mortality reduces the size of the flock.

When the rooster mates with the hen, sperm enter the hen's oviduct and are stored within sperm storage glands. These glands can store more than half a million sperm, and sperm can remain viable for up to 3 weeks. However, a hen will have maximum fertility for only about 3 to 4 days after one mating. Therefore, the male-to-female ratio in a flock must be enough to ensure mating of every hen every 3 days or so. To maintain fertility, younger roosters may be introduced as the flock ages.

Eggs are collected a minimum of twice a day, and usually more frequently. Cracked or dirty eggs are separated, as they are not suitable for hatching. Undersized, oversized or double-yolk eggs are also unsuitable. The eggs might be disinfected by fumigation, are packed in 'flats' or trays, placed in wheeled trolleys, and stored in a cool (15-18 °C) climate-controlled area. The egg packing room and storage rooms are kept segrated to reduce contamination. The trolleys are delivered by truck to a hatchery perhaps twice a week.

At the end of the production cycle, the birds are called "spent fowl". Disposal of spent fowl may be a problem as consumer demand for them is poor.

Hatcheries

Five-day-old broiler strain Cornish-Rock chicks.

Hatcheries take the fertilized eggs, incubate them, and produce day old broiler chicks.

Incubation takes about 21 days, and is often a two-step process. Initial incubation is done in machines known as *setters*. A modern setter is the size of a large room, with a central corridor and racks on either side. Eggs are held relatively tightly (large end up) in trays, which are stored in the racks. Inside the setter, temperature and humidity are closely maintained. Blowers or fans circulate air to ensure uniform temperature, and heating or cooling is applied as needed by automated controls. The racks pivot or tilt from side to side, usually on an hourly basis. As an example, one commercial machine can hold up to 124,416 eggs and measures about 4.7 metres wide by 7.2 metres deep. Setters often hold more than one hatch, on a staggered hatch-day basis, and operate continuously. The setter phase lasts about 18 days.

On or about day 18, the eggs are removed from the setters and transferred to hatchers. These machines are similar to setters, but have larger flat-bottom trays, so the eggs can rest on their sides, and newly hatched chicks can walk. Having a separate machine helps keep hatching debris out of the setter. The environmental conditions in the hatcher are optimized to help the chicks hatch. As a commercial example, a large hatcher has capacity for 15,840 eggs, and measures about 3.3 metres by 1.8 metres.

Some incubators are single-stage (combining setter and hatcher funcations), and entire trolleys of eggs can be rolled in at one time. One advantage of single-stage machines is that they are thoroughly cleaned after each hatch, whereas a setter is rarely shutdown for cleaning. The single-stage environment can be adjusted for eggs from different producing flocks, and for each step of the hatch cycle. The setter environment is often a compromise as different egg batches are in the machine at one time.

On hatch day (day 21), the trays are removed ("pulled") from the hatchers, and then the chicks are removed from the trays. Chicks are inspected, with the sickly ones being disposed of. Chicks may be by vaccinated, sorted by sex, counted, and placed in chick boxes. Stacks of chick boxes are loaded into trucks for transport, and arrive at the broiler farm on the same day. Specialized climate-controlled trucks are typically used, depending on climate and transport distance.

Chick sexing is optionally done to improve uniformity – since males grow faster, the weights are more uniform if males and females are raised separately. The birds are bred so that males and females have unique feather patterns or color differences. Unlike egg-laying poultry, males are not culled.

Typical hatchability rate in Canada in 2011 was 82.2%. (i.e. 82.2% of eggs set for incubation produced a saleable chick). A UK source estimates 90% hatchability.

Broiler Farms

The chicks are delivered to the actual broiler *Grow-Out* farms. In the US, houses may be up to 60' x 600' (36000 sq.ft.). One 2006 magazine survey reported a desired 67 foot wide house, with the average 'standard' new house being 45' x 493', with largest being 60' x 504'. One farm complex may have several houses.

Broiler chickens in a farm.

In Mississippi, typical farms now have four to six houses with 25,000 birds per house. One full-time worker might manage three houses. On average, a new broiler house is about 500 feet long by 44 feet wide and costs about $200,000 equipped.

When the birds are full-grown, they are caught (perhaps with a chicken harvester) placed in crates, and transported by truck to a processing plant.

Broiler chickens kept outside near a chicken shop in India.

Because of their efficient meat conversion, broiler chickens are also popular in small family farms in rural communities, where a family will raise a small flock of broilers.

Processing Plants

When the birds are large enough, they are shipped to processing plants for slaughter. When chickens arrive at the processor they go through the following sequence:

- Removed from transport cages.

- Hung by the legs on a shackle, mounted on a conveyor chain.

- Stunned using an electrically charged water bath.

- Killed by cutting the blood vessels in the neck.

- Bled so that most blood has left the carcass.

- Scalded to soften the attachment of the feathers.

- Plucked to remove the feathers.

- Head removed.

- Hock cutting to remove the feet.

- Rehung in the evisceration room.

- Gutted or eviscerated to remove the internal organs.

- Washed to remove blood and soiling from the carcass.

- Chilled to prevent bacterial spoiling (They go through a chiller which takes approximately 2 hours to go through. The chiller generally holds thousands of gallons of water kept below 40 degrees Fahrenheit).

- Drained to allow excess water to drip off the carcass.

- Weighing.

- Cut selection to divide the carcass into desired portion (breast, drumsticks etc).

- Packed (for example in plastic bags) to protect carcasses or cuts.

- Chilled or frozen for preservation.

Further Processing plants carry out operations such as cutting and deboning. Previously the conveyor belts carrying live chickens generally ran at a maximum of 140 chickens per minute, but the maximum speed has been increased to 175 birds/minute. Once the dead birds arrive in the evisceration room (usually dropped down a chute after the feet are removed), they are hung again on shackles much the same way as they were when they were alive.

PEACOCK FARMING

Peacock farming is not a new business idea. From the ancient time, people like bird's egg and meat on their table. People raise various types of birds for the purpose of producing meat, eggs, and also for their beauty. Raising various types of poultry birds are one of the best and lucrative business ventures. Along with profit, people usually keep various types of birds for the purpose of producing fresh food. Some people also raise birds as a hobby and for entertaining purpose.

Generally two legged creatures who have feathers are called birds. In other words, birds are the back boned creatures who have the ability to fly and to control body temperature. And the birds which are raised commercially for the purpose of producing meat, egg or features are known as poultry. Peacock is also a poultry bird, like ducks, chicken, pigeon, turkey etc. Poultry farming business is very profitable and becoming popular day by day.

Peacock in the Poultry Industry

Duck, chicken, quail etc. are most common and popular poultry birds. Most of the people consider chickens as poultry, but that's not true. Actually all types of birds raised for their meat or egg production are called poultry. Most of the people in rural areas raise some birds, such as pigeon, ducks, chicken, swan, quail etc. for the purpose of producing meat or eggs. And all these birds are poultry. Peacocks are also among those poultry birds.

Nowadays peacock farming is considered as a profitable poultry farming business. You can raise peacock commercially in both rural and urban areas. Peacock farming can play an important role for reducing unemployment problem and help to create a new way of earnings. And commercial peacock farming business can help to develop the socioeconomic condition of a country or nation.

Features of Peacock

The peacock has iridescent blue-green or green colored plumage. Head, neck, van and throat protector is little blue colored. Both peacock and peahens have crest atop the head. Their legs are reddish. Peacocks have very attractive tail and they use their beautiful tail for attracting the peahens. You can see 'eyes' on their tail, when they fan their tail. Peacocks dance raising their tail plumage to attract peahens, which is very impressive and mind blowing. Peacocks always love to stay in a group. Usually you will see 1 peacock with 5 to 6 peahens in a group. In the wild peacock move over the ground for a short period, most of the time they stay upon the large trees. Grains, insects, snakes, small fish, frogs, etc are their favorite food. Peahens lay about 3 to 5 eggs by making a small hole in the ground. The eggs are off white colored. And it takes about 27 to 30 days for hatching their eggs. Average life expectancy of peafowl is around 35 years. They reach sexual maturity at their 3 to 4 years of age.

Domestic Peacock Farming

If you want to raise peafowl on your property, then you have to feed them nutritious foods and make a suitable environment for them. Peafowls are omnivorous, their regular includes wheat, rice, vegetables, seeds, insects, snails etc. Papaya, melon and other fruits are also their favorite food. Along with providing nutritious food, you have to supply sufficient amount of fresh and clean water daily.

Breeds

There are many peafowl breeds available around the world. Usually peafowls have three species and rest are the hybirds or crosses. Some common peafowl breeds are Blue Indian, Black Shoulder, Bufford Bronze, Charcoal, Cameo, Green Peafowl (Java Green), Opal, Oaten, Spalding, Purple, White, Pied etc. Among these breeds, Blue Indian are mostly raised peafowl. Most of the people like this breed mainly for their mild and timid nature.

Behavior

Peafowls do not bathe in water, like other pheasants. They clean themselves by vigorously rubbing their plumage in dust and dry soil. They love to have dust baths. Indian blue peafowls are the mostly raised domestic species and very popular for their mild and timid nature. Green peafowls are aggressive in nature and they are less common as domestic bird.

Housing

In case of domestic peafowl farming, you have to ensure adequate housing and good environment. Usually 100 square feet space required for each bird. Adjust the housing space according to the length of peacock's trains. Generally peacock trains can be over 5 feet long. Their shed must have to be at least 8 feet tall. You can use chicken wire for the walls and roof. Provide a wooden shelter similar to a small shed or barn. You can bed this area with straw and keep a warming light inside. Provide large shed or barn, if you have nesting peahens. Use hanging dishes inside the shed for providing foods, and keep the water elevated so that you do not get droppings in it. You will need to cover their pen, because peafowls are great flyers. Usually peafowl like to roost at night. You have to make suitable roosting place for them inside the house. Ensure well ventilation system. Make the house suitable enough for preventing all types of predators, such as raccoon wild dogs or foxes.

Feeding

Peafowls are omnivorous. They eat almost everything whether any plant or animal matter. Wild figs, berries and nuts are favorite foods in season, but seeds, grain and leaves are staple items all year. They also love to catch and eat mice, frogs, small snakes, various types of small mammals, various types of insects such as ants, grasshoppers, termites etc. Peafowl don't drink water frequently like other domestic poultry birds. Usually they drink water during their midday rest and while eating in late afternoon. They drink again just before going to roost. You can also feed your peafowl a wide range of crops and fruits such as bananas, chilly, groundnut, paddy, tomato etc. If you have any peachicks, you can feed them turkey starter feed, boiled chicken eggs etc.

Breeding

Usually peafowl's breeding season starts with the monsoon. There are some advantages of mating during this season. Peachicks benefit from the abundance of food resulting from the rains. Peacocks show their trains and dance for attracting the peahens. Each peacock mate with 5 to 6 peahens in his group during this season. After successful mating, peahens lay eggs. Usually it takes about 27 to 30 days to hatch their eggs.

Marketing

Peacock farming is well-spread throughout the world and it is considered as a very lucrative business. Most of the peacock farmer raise them for the purpose of producing their colorful feathers. Peacock feathers have a huge demand around the world, especially in the European countries. As far as we have learned, in some places price of a single feather ranges between 1 to 5 euros. Although the price depend on the size. Peacocks change their plumage once a year and you can get about 150 to 200 feathers from and adult bird every year. If you have any local customers, then you can also sell eggs and live birds. Peacock meat is also popular in some areas. So if you want to start commercial peacock farming business, consider about marketing your products first.

4

Poultry Management

The husbandry practices and production techniques that help to increase the efficiency of poultry production is referred to as poultry management. The main practices which fall under poultry management are brooding management, poultry housing and management, poultry rearing, chick culling, forced molting, etc. This chapter has been carefully written to provide an easy understanding of these methods and techniques of poultry management.

Poultry management usually refers to the husbandry practices or production techniques that help to maximize the efficiency of production. Sound management practices are very essential to optimize production. Scientific poultry management aims at maximizing returns with minimum investment.

Brooder Management

Brooder house: Brooder house should be draft-free, rain-proof and protected against predators. Brooding pens should have windows with wire mesh for adequate ventilation. Too dusty environment irritates the respiratory tract of the chicks. Besides dust is one of the vehicles of transmission of diseases. Too much moisture causes ammonia fumes which irritate the respiratory tract and eyes. Good ventilation provides a comfortable environment without draft.

Sanitation and Hygiene

All movable equipments like feeders, waterers and hovers should be removed from the house, cleaned and disinfected. All litters are to be scraped and removed. The interior as well as exterior of the house should be cleaned under pressure. The house should be disinfected with any commercial disinfectant solution at the recommended concentration. Insecticide should be sprayed to avoid insect threat. Malathion spray/blow lamping or both can be used to control ticks and mites. New litter should be spread after each cleaning. The insecticides if necessary should be mixed with litter at recommended doses.

Litter

Suitable litter material like saw dust and paddy husk should be spread to a length of

5 cm depending upon their availability and cost. Mouldy material should not be used. The litter should be stirred at frequent intervals to prevent caking. Wet litters if any should be removed immediately and replaced by dry new litter. This prevents ammoniacal odour.

Brooding Temperature

Heating is very much essential to provide right temperature in the brooder house. Too high or too low a temperature slows down growth and causes mortality. During the first week the temperature should be 95 °F (35 °C) which may be reduced by 5 °F per week during each successive week till 70 °F (21·1 °C). The brooder should be switched on for at least 24 hours before the chicks arrive. As a rule of thumb the temperature inside the brooder house should be approximately 20°F (-6·7 °C) below the brooder temperature Hanging of a maximum and minimum thermometer in each house is recommended to have a guide to control over the differences in the house temperature. The behavior of chicks provides better indication of whether they are getting the desired amount of heat. When the temperature is less than required, the chicks try to get closer to the source of heat and huddle down under the brooder. When the temperature is too high, the chicks will get away from the source of heat and may even pant or gasp. When temperature is right, the chicks will be found evenly scattered. In hot weather, brooders are not necessary after the chicks are about 3 weeks old. Several devices can be used for providing artificial heat. Hover type electric brooders are by far the most common and practical these days. The temperature in these brooders is thermostatically controlled. Many a times the heat in the brooder house is provided by use of electric bulbs of different intensities. Regulation of temperature in such cases is difficult although not impossible. Infrared lamps are also very good for brooding. The height and number of infra-red lamps can be adjusted as per temperature requirement in the brooder house.

Brooder

Brooder Space

Brooder space of 7 to 10 sq inch (45-65 cm2) is recommended per chick. Thus a 1·80 m hover can hold 500 chicks. When small pens are used for brooding, dimension of the house must be taken into consideration as overcrowding results in starve-outs, culls and increase in disease problems.

Brooder Guard

To prevent the straying of baby chicks from the source of heat, hover guards are placed 1·05 to 1·50 m from the edge of hover. Hover guard is not necessary after 1 week.

Floor Space

Floor space of 0·05 m² should be provided per chick to start with, which should be increased by 0·05 m² after every 4 weeks until the pullets are about 20 weeks of age. For broilers at least 0·1 m² of floor space for female chicks and 0·15 m² for male chicks should be provided till 8 weeks of age. Raising broiler pullets and cockerel chicks in the separate pens may be beneficial.

Water Space

Plentiful of clean and fresh water is very much essential. A provision of 50 linear cm of water space per 100 chicks for first two weeks has to be increased to 152-190 linear cm at 6 to 8 weeks. When changing from chick fountain to water trough the fountains are to be left in for several days till the chicks have located the new water source. Height of the waterers should be maintained at 2·5 cm above the back height of the chicks to reduce spoilage. Antibiotics or other stress medications may be added to water if desired. All waterers should be cleaned daily. It may be desirable to hold a few chicks one at a time and teach them to drink.

FREE RANGE

Commercial free range hens in Scotland.

Free range denotes a method of farming husbandry where the animals, for at least part of the day, can roam freely outdoors, rather than being confined in an enclosure for 24 hours each day. On many farms, the outdoors ranging area is fenced, thereby technically making this an enclosure, however, free range systems usually offer the opportunity for the extensive locomotion and sunlight that is otherwise prevented by indoor housing systems. Free range may apply to meat, eggs or dairy farming.

The term is used in two senses that do not overlap completely: as a farmer-centric description of husbandry methods, and as a consumer-centric description of them. There is a diet where the practitioner only eats meat from free-range sources called ethical omnivorism.

In ranching, free-range livestock are permitted to roam without being fenced in, as opposed to fenced-in pastures. In many agriculture-based economies, free-range livestock are quite common.

A small flock of mixed free-range chickens being fed outdoors.

If one allows "free range" to include "herding", free range was a typical husbandry method at least until the development of barbed wire and chicken wire. The generally poor understanding of nutrition and diseases before the twentieth century made it difficult to raise many livestock species without giving them access to a varied diet, and the labor of keeping livestock in confinement and carrying all their feed to them was prohibitive except for high-profit animals such as dairy cattle.

In the case of poultry, free range was the dominant system until the discovery of vitamins A and D in the 1920s, which allowed confinement to be practised successfully on a commercial scale. Before that, green feed and sunshine (for the vitamin D) were necessary to provide the necessary vitamin content. Some large commercial breeding flocks were reared on pasture into the 1950s. Nutritional science resulted in the increased use of confinement for other livestock species in much the same way.

Free range ducks in Hainan Province, China.

In the United States, the USDA free range regulations currently apply only to poultry and indicate that the animal has been allowed access to the outside. The USDA regulations do not specify the quality or size of the outside range nor the duration of time an animal must have access to the outside.

The term "free range" is mainly used as a marketing term rather than a husbandry term, meaning something on the order of, "low stocking density," "pasture-raised," "grass-fed," "old-fashioned," "humanely raised," etc.

There have been proposals to regulate USDA labeling of products as free range within the United States. As of 2017 what constitutes raising an animal "free range" is almost entirely decided by the producer of that product, and is frequently inconsistent with consumer ideas of what the term means.

Free-range Poultry

Free range meat chickens seek shade on a U.S. farm.

In poultry-keeping, "free range" is widely confused with yarding, which means keeping poultry in fenced yards. Yarding, as well as floorless portable chicken pens ("chicken tractors") may have some of the benefits of free-range livestock but, in reality, the methods have little in common with the free-range method.

A behavioral definition of free range is perhaps the most useful: "chickens kept with a fence that restricts their movements very little." This has practical implications. For example, according to Jull, "The most effective measure of preventing cannibalism seems to be to give the birds good grass range." De-beaking was invented to prevent cannibalism for birds not on free range, and the need for de-beaking can be seen as a litmus test for whether the chickens' environment is sufficiently "free-range-like."

The U.S. Department of Agriculture Food Safety and Inspection Service (FSIS) requires that chickens raised for their meat have access to the outside in order to receive the free-range certification. There is no requirement for access to pasture, and there may be access to only dirt or gravel. Free-range chicken eggs, however, have no legal definition in the United States. Likewise, free-range egg producers have no common standard on what the term means.

The broadness of "free range" in the U.S. has caused some people to look for alternative terms. "Pastured poultry" is a term promoted by farmer/author Joel Salatin for broiler chickens raised on grass pasture for all of their lives except for the initial brooding period. The Pastured Poultry concept is promoted by the American Pastured Poultry Producers' Association (APPPA), an organization of farmers raising their poultry using Salatin's principles.

Free-range Livestock

Traditional American usage equates "free range" with "unfenced," and with the implication that there was no herdsman keeping them together or managing them in any way. Legally, a free-range jurisdiction allowed livestock (perhaps only of a few named species) to run free, and the owner was not liable for any damage they caused. In such jurisdictions, people who wished to avoid damage by livestock had to fence them out; in others, the owners had to fence them in.

The USDA has no specific definition for "free-range" beef, pork, and other non-poultry products. All USDA definitions of "free-range" refer specifically to poultry.

In a December 2002 Federal Register notice and request for comments, USDA's Agricultural Marketing Service proposed "minimum requirements for livestock and meat industry production/marketing claims". Many industry claim categories are included in the notice, including breed claims, antibiotic claims, and grain fed claims. "Free Range, Free Roaming, or Pasture Raised" would be defined as "livestock that have had continuous and unconfined access to pasture throughout their life cycle" with an exception for swine ("continuous access to pasture for at least 80% of their production cycle"). In a May 2006 Federal Register notice, the agency presented a summary and its responses to comments received in the 2002 notice, but only for the category "grass (forage) fed" which the agency stated was to be a category separate from "free range."

Small-scale free range farming in the Northern Black Forest.

The European Union regulates marketing standards for egg farming which specifies the following (cumulative) minimum conditions for the free-range method:

- Hens have continuous daytime access to open-air runs, except in the case of temporary restrictions imposed by veterinary authorities,

- The open-air runs to which hens have access is mainly covered with vegetation and not used for other purposes except for orchards, woodland and livestock grazing if the latter is authorised by the competent authorities,

- The maximum stocking density is not greater than 2500 hens per hectare of ground available to the hens or one hen per 4m² at all times and the runs are not extending beyond a radius of 150 m from the nearest pophole of the building; an extension of up to 350 m from the nearest pophole of the building is permissible provided that a sufficient number of shelters and drinking troughs within the meaning of that provision are evenly distributed throughout the whole open-air run with at least four shelters per hectare.

Free range geese in Germany.

Otherwise, egg farming in EU is classified into 4 categories: Organic (ecological), Free Range, Barn, and Cages.) The mandatory labelling on the egg shells attributes a number (which is the first digit on the label) to each of these categories: 0 for Organic, 1 for Free Range, 2 for Barn and 3 for Cages.

There are EU regulations about what free-range means for laying hens and broilers (meat chickens) as indicated above. However, there are no EU regulations for free-range pork, so pigs could be indoors for some of their lives. In order to be classified as free-range, animals must have access to the outdoors for at least part of their lives.

United Kingdom

Free range pigs in England.

Pigs: Free-range pregnant sows are kept in groups and they are often provided with straw for bedding, rooting and chewing. Around 40% of UK sows are kept free-range outdoors and farrow in huts on their range.

Egg laying hens: Cage-free egg production includes barn, free-range and organic systems. In the UK, free-range systems are the most popular of the non-cage alternatives, accounting for around 57% of all eggs, compared to 2% in barns and 2% organic. In free-range systems, hens are housed to a similar standard as the barn or aviary.

Free-range rearing of pullets: Free range rearing of pullets for egg-laying is now being pioneered in the UK by various poultry rearing farms. In these systems, the pullets are allowed outside from as young as 4 weeks of age, rather than the conventional systems where the pullets are reared in barns and allowed out at 16 weeks of age.

Meat chickens: Free-range broilers are reared for meat and are allowed access to an outdoor range for at least 8 hours each day. Free-range broiler systems use slower-growing breeds of chicken to improve welfare, meaning they reach slaughter weight at 16 weeks of age rather than 5–6 weeks of age in standard rearing systems.

Turkeys: Free-range turkeys have continuous access to an outdoor range during the daytime. The range should be largely covered in vegetation and allow more space. Access to fresh air and daylight means better eye and respiratory health. The turkeys are able to exercise and exhibit natural behavior resulting in stronger, healthier legs. Free-range systems often use slower-growing breeds of turkey.

Free range Dairy: In recent years the free range dairy scheme has become more prevalent. Farms supplying milk under the free range diary brand abide by the pasture promise, meaning the cows will have access to pasture land to graze for a minimum of 180 days and nights a year. There is evidence to suggest that milk from grass contains higher levels of fats such as omega-3 and conjugated linoleic acid (CLA). Additionally free range dairy is giving consumers more choice as to where their milk comes from. Free range dairy provides the consumer with reassurance that the milk they drink has come from cows with the freedom to roam and can graze in their natural habitat.

YARDING

During the daytime, the doors are left open for these chickens to choose
whether to be in the yard or coop.

In poultry keeping, yarding is the practice of providing the poultry with a fenced yard in addition to a poultry house. Movable yarding is a form of managed intensive grazing.

Yarding is often confused with free range. The distinction is that free-range poultry are either totally unfenced, or the fence is so distant that it has little influence on their freedom of movement.

Before the discovery of vitamins A and D in the 1920s, green feed and sunshine were essential to the health of poultry. Vitamin D was synthesized from sunlight on the skin (as with humans), while Vitamin A was obtained through green forage plants such as grass. Yards small enough to be fenced economically were soon stripped of palatable green forage and become barren. This is followed by a build-up of manure, parasites, and other pathogens.

Free range husbandry was the most common method in these early days. Most farms had only a small free-range barnyard flock. Larger flocks were kept in small houses build on skids, which were dragged periodically to a fresh piece of ground. This method is similar to the modern practice of pastured poultry.

Experts of the day estimated the sustainable level to be about fifty hens per acre (80 m^2 per hen), with one hundred hens per acre (40 m^2 per hen) as an absolute upper limit if special care was taken. These levels are sustainable in the sense that the turf can make use of the nutrients in the manure left behind by the chickens, and in the sense that, at this stocking density, the chickens will not completely destroy the turf through scratching.

At the Oregon Station on clay soil it was found that the day droppings from 200 laying hens on an acre 20 m^2 per hen in four years made the soil too rich for the successful growth of cereal crops where cropping the ground was done every other year. The night droppings were put on other land. If the soil contains too much manure for the crops it is safe to assume that it is not in the best condition for poultry. Sooner or later it is bound to show not only in a failure of grain crops but in failure of poultry crops. For a permanent system under average conditions of soil and climate the following points are suggested for consideration:

- Maximum number of fowls per acre: 100 laying hens 40 m^2 per hen.

- Disposing of the night droppings on other land.

- Dividing the ground into at least two divisions or yards, and growing a crop on each yard at least every other year. In topics where crops may be grown every year the number of fowls may be increased.

- Growing crops that will use up the maximum amount of manure.

- Keeping the ground vacant of chickens at least six months in the year.

- Thorough underdrainage, where necessary, to carry off surplus water.

It is not assumed that as many as 500 hens may not be profitably kept on an acre 8 m² per hen for a few years under favorable conditions. It has been done, but it is a different matter when it is planned to make a permanent business of it.

Because fifty hens per acre represents 800 square feet (74 m²) per hen (80 m² per hen), while the density inside the house at the time was normally four square feet per hen (0.4 m² per hen), this required that the yard be 200 times wider than the house, assuming a yard on one side of the house. That is, a house 20 feet (6 m) wide required a yard 4,000 feet (1,220 m) wide to provide the necessary area. This would normally be provided as two yards, one on either side of the house, each 2,000 feet (610 m) wide. In reality, such yards are expensive to fence, and the chickens spend most of their time on the portion closest to the house, so sustainability was never achieved in practice except with portable houses, which were moved periodically to fresh ground. Yarded operations were operated with unsustainably small yards that were quickly denuded and which received excessive levels of manure.

The use of multiple yards, frequent plowing, and liberal use of lime would allow higher stocking levels to be used, since plowing and liming would allow much of the nitrogen to escape from the soil.

The following is typical advice for the successful use of yards in the Thirties and Forties:

All poultrymen should realize that there are no known substitutes for sunshine and young green grass in keeping poultry in the best possible state of health and in promoting growth and maintaining egg production. Where sunshine and green grass cannot be provided, as in the case of birds kept in strict confinement, the best possible substitutes must be provided. In the case of most farm and many commercial flocks, however, the growing stock is reared on range, and the adult birds are given yards or allowed to roam at will.

If the staggering losses among growing chicks and laying birds that occur annually are to be reduced materially, better methods of flock management must be employed. The losses from mortality are due largely to internal parasites and diseases of one kind or another. Bare ground over which the chickens have run for some time, mud puddles, and stagnant water are the chief sources of the spread of diseases, most of which are filth borne.

The mortality that usually occurs in growing and adult stock may be materially reduced by providing the birds with an alternate yarding system. Probably the best arrangement is to provide each colony brooder and each laying house with three yards (3 m) which the birds would be allowed to use every 3 or 4 weeks. By alternating the birds in the yards every 3 or 4 weeks each yard is kept reasonably sanitary, especially if the soil in the immediate vicinity of the house is cultivated and treated with lime, and young green grass is available for the birds throughout the season. The importance of clean range for both birds and adult stock cannot be emphasized too strongly. For adult stock a good grass sward can be maintained on fertile soil, allowing about 200 birds to the acre 20 m² per hen.

Nutritional advances increasingly turned yarding into a liability, and it fell out of favor. Free range continued to be used, especially for breeding flocks and for pullets before they reached laying age, because of the lower rate of disease and greater overall health of grass-reared chickens. Breeding flocks (which lay eggs destined for incubation) are always given a better diet than flocks laying table eggs, since a diet that will produce table eggs cheaply will not provide eggs that hatch well. For some time after confined laying flocks produced table eggs satisfactorily, breeding flocks benefited from free range.

In Britain, Geoffrey Sykes developed a new yarding system in the Fifties. This used a small yard covered with a thick layer of straw, with more straw added frequently. He also recommended that shade and a windbreak be provided by a solid fence around the yard, or by other means, such as rows of haybales. Once a year, the old straw was removed by a front-end loader or similar machinery. This method eliminated mud and pathogens. It was later forgotten because the industry moved to high-density confinement before the method was widely established.

Today, commercial poultry producers generally call yarding free range on their labels. This conflation of two very different techniques has led to confusion. The vast majority of "free-range" operations are really yarded.

Pastured poultry, as promoted by the APPPA, the American Pastured Poultry Producers Association, and author/farmer Joel Salatin, takes a different approach, attempting to achieve the benefits of free range while using penning or yarding. The key element of Pastured poultry is the use of portable housing and the optional use of portable electric fencing. By moving the house and yard frequently, perhaps daily, all the disadvantages of permanent yards are eliminated.

BATTERY CAGE

Battery cages are a housing system used for various animal production methods, but primarily for egg-laying hens. The name arises from the arrangement of rows and columns of identical cages connected together, in a unit, as in an artillery battery. Although the term is usually applied to poultry farming, similar cage systems are used for other animals. Battery cages have generated controversy between advocates for animal rights and industrial producers.

Battery Cages in Practice

Battery cages are the predominant form of housing for laying hens worldwide. They reduce aggression and cannibalism among hens, but are barren, restrict movement, prevent many natural behaviors, and increase rates of osteoporosis. As of 2014, approximately 95% of eggs in the US were produced in battery cages. In the UK, statistics from the Department for the Environment, Food and Rural Affairs (Defra) indicate that 50% of eggs produced in the UK throughout 2010 were from cages (45% from free-range, 5% from barns).

Battery cages also used for mink, rabbit, chinchilla and fox in fur farming, and most recently for the Asian palm civet for kopi luwak production of coffee.

Battery cages for mink reared for their fur.

Battery cages for civets reared for kopi luwak (coffee) production.

An early reference to battery cages appears in Milton Arndt's 1931 book, *Battery Brooding*, where he reports that his cage flock was healthier and had higher egg production than his conventional flock. At this early date, battery cages already had the sloped floor that allowed eggs to roll to the front of the cage, where they were easily collected by the farmer and out of the hens' reach. Arndt also mentions the use of conveyor belts under the cages to remove manure, which provides better air control quality and reduces fly breeding.

Original battery cages extended the technology used in battery brooders, which were cages with a wire mesh floor and integral heating elements for brooding chicks. The wire floor allowed the manure to pass through, removing it from the chicks' environment and reducing the risk of manure-borne diseases.

Early battery cages were often used for selecting hens based on performance, since it is easy to track how many eggs each hen is laying if only one hen is placed in a cage. Later, this was combined with artificial insemination, giving a technique where each egg's parentage is known. This method is still used today.

A chicken coop from the 1950s.

Early reports from Arndt about battery cages were enthusiastic. Arndt reported:

> "This form of battery is coming into widespread use throughout the country and apparently is solving a number of the troubles encountered with laying hens in the regular laying house on the floor. The first edition of book which spoke of experimental work with 220 pullets which were retained for one year in individual cages. At the end of this year it was found that the birds confined in the batteries outlaid considerably the same size flock in the regular houses. The birds consume less feed than those on the floor and this coupled with the increased production made them more profitable than the same number of pullets in the laying house".

A number of progressive poultrymen from all over the United States and some in foreign countries cooperated with me in carrying on experimental work with this type of battery and each and every one of them were very well satisfied with the results obtained. In fact, a number of them have since placed their entire laying flocks in individual hen batteries.

The use of laying batteries increased gradually, becoming the dominant method somewhat before the integration of the egg industry in the 1960s. The practice of battery cages was criticized in Ruth Harrison's landmark book *Animal Machines*, published in 1964.

A simple battery cage system with no conveyors for feed or eggs.

In 1990, North and Bell reported that 75% of all commercial layers in the world and 95% in the United States were kept in cages.

By all accounts, a caged layer facility is more expensive to build than high-density floor confinement, but can be cheaper to operate if designed to minimize labor.

North and Bell report the following economic advantages to laying cages:

1. It is easier to care for the pullets; no birds are underfoot.

2. Floor eggs are eliminated.

3. Eggs are cleaner.

4. Culling is expedited.

5. In most instances, less feed is required to produce a dozen eggs.

6. Broodiness is eliminated.

7. More pullets may be housed in a given house floor space.

8. Internal parasites are eliminated.

9. Labor requirements are generally much reduced.

They also cite disadvantages to cages:

1. The handling of manure may be a problem.

2. Generally, flies become a greater nuisance.

3. The investment per pullet may be higher than in the case of floor operations.

4. There is a slightly higher percentage of blood spots in the eggs.

5. The bones are more fragile and processors often discount the fowl price.

Disadvantages 1 and 2 can be eliminated by manure conveyors, but some industrial systems do not feature manure conveyors.

Welfare Concerns

There are several welfare concerns regarding the battery cage system of housing and husbandry. These are presented below in the approximate chronological order they would influence the hens.

Chick Culling

Due to modern selective breeding, laying hen strains are different from those of meat production strains. Male birds of the laying strains do not lay eggs and are unsuitable for meat production, therefore, they are culled soon after being sexed, often on the day of hatching. Methods of culling include cervical dislocation, asphyxiation by carbon dioxide and maceration using a high speed grinder.

Animal rights groups have used videos of live chicks being placed into macerators as evidence of cruelty in the egg production industry. Maceration, together with cervical dislocation and asphyxiation by carbon dioxide, are all considered acceptable methods of euthanasia by the American Veterinary Medical Association. Consumers may also be appalled simply by the death of animals that are not subsequently eaten.

Beak-trimming

To reduce the harmful effects of feather pecking, cannibalism and vent pecking, most chicks eventually going into battery cages are beak-trimmed. This is often performed on the first day after hatching, simultaneously with sexing and receiving vaccinations. Beak-trimming is a procedure considered by many scientists to cause acute pain and distress with possible chronic pain; it is practised on chicks for all types of housing systems, not only battery cages.

Cage Size

Battery cage.

At approximately 16 weeks of age, pullets (hens which have not yet started to lay) are placed into cages. In countries with relevant legislation, floor space for battery cages ranges upwards from 300 cm² per bird. EU standards in 2003 called for at least

550 cm² per hen. In the US, the current recommendation by the United Egg Producers is 67 to 86 in² (430 to 560 cm²) per bird. The space available to each hen in a battery cage has often been described as less than the size of a sheet of A4 paper (624 cm²). Others have commented that a typical cage is about the size of a filing cabinet drawer and holds eight to 10 hens.

Behavioral studies showed that when turning, hens used 540 to 1006 cm², when stretching wings 653 to 1118 cm², when wing flapping 860 to 1980 cm², when feather ruffling 676 to 1604 cm², when preening 814 to 1240 cm², and when ground scratching 540 to 1005 cm². A space allowance of 550 cm² would prevent hens in battery cages from performing these behaviors without touching another hen. Animal welfare scientists have been critical of battery cages because of these space restrictions and it is widely considered that hens suffer boredom and frustration when unable to perform these behaviors. Spatial restriction can lead to a wide range of abnormal behaviors, some of which are injurious to the hens or their cagemates.

Light Manipulation

Battery cages - note the low light intensity beyond range of the camera flashgun.

To reduce the harmful effects of feather pecking, cannibalism and vent-pecking, hens in battery cages (and other housing systems) are often kept at low light intensities (e.g. less than 10 lux). Low light intensities may be associated with welfare costs to the hens as they prefer to eat in brightly lit environments and prefer brightly lit areas for active behavior but dim (less than 10 lux) for inactive behavior. Dimming the lights can also cause problems when the intensity is then abruptly increased temporarily to inspect the hens; this has been associated as a risk factor of increased feather pecking and the birds can become frightened resulting in panic-type ("hysteria") reactions which can increase the risk of injury.

Being indoors, hens in battery cages do not see sunlight. Whilst there is no scientific evidence for this being a welfare problem, some animal advocates indicate it is a concern. Furnished cages and some other non-cage indoor systems would also prevent hens seeing natural light throughout their lives.

Osteoporosis

Several studies have indicated that toward the end of the laying phase (approximately 72 weeks of age), a combination of high calcium demand for egg production and a lack of exercise can lead to osteoporosis. This can occur in all housing systems for egg laying hens, but is particularly prevalent in battery cage systems where it has sometimes been called 'cage layer osteoporosis'. Osteoporosis leads to the skeleton becoming fragile and an increased risk of bone breakage, particularly in the legs and keel bone. Fractures may occur whilst the hens are in the cage and these are usually discovered at depopulation as old, healed breaks, or they might be fresh breaks which occurred during the process of depopulation. One study showed that 24.6% of hens from battery cages had recent keel fractures whereas hens in furnished cages, barn and free-range had 3.6%, 1.2% and 1.3% respectively. However, hens from battery cages experienced fewer old breaks (17.7%) compared to hens in barn (69.1%), free-range (59.8%) and furnished cages (31.7%).

Forced Moulting

Flocks are sometimes force moulted, rather than being slaughtered, to reinvigorate egg-laying. This involves complete withdrawal of food (and sometimes water) for 7 to 14 days or sufficiently long to cause a body weight loss of 25 to 35%. This stimulates the hen to lose her feathers, but also reinvigorates egg-production. Some flocks may be force moulted several times. In 2003, more than 75% of all flocks were moulted in the US. This temporary starving of the hens is seen as inhumane and is the main point of objection by critics and opponents of the practice. The alternative most often employed is to slaughter the hens instead of moulting them.

Improving Welfare for Egg Producing Hens

The Scientific Veterinary Committee of the European Commission stated that "enriched cages and well designed non-cage systems have already been shown to have a number of welfare advantages over battery systems in their present form". Supporters of battery husbandry contend that alternative systems such as free range also have welfare problems, such as increases in cannibalism, feather pecking and vent pecking. A recent review of welfare in battery cages made the point that such welfare issues are problems of management, unlike the issues of behavioral deprivation, which are inherent in a system that keeps hens in such cramped and barren conditions. Free range egg producers can limit or eliminate injurious pecking, particularly feather pecking, through such strategies as providing environmental enrichment, feeding mash instead of pellets, keeping roosters in with the hens, and arranging nest boxes so hens are not exposed to each other's vents; similar strategies are more restricted or impossible in battery cages.

Benefits of using Poultry Battery Cages

In the modern day chicken rearing, battery systems have certain beneficial impact on poultry farming in the following seven ways.

Battery Cages Boosts the Health of the Chickens

Many poultry farmers do not get healthy results from their chicken rearing because they have ignored the use of battery cages. Battery cages are very good means of accommodating your chickens in healthy conditions that will boost their reproduction.

Original battery cages are not mere cages with wire mesh floor and essential heating elements for brooding chickens. Quality and superior battery cages are designed to outweigh an ordinary wire mesh cage. The wire floor is designed to allow manure go through it, taking it away from the chickens surroundings and eventually minimizing the risks of manure-borne diseases. This will keep the chickens healthy and make chicken rearing a task worthwhile for the poultry farmer.

Increases Egg Production

A chicken that is healthy will significantly become more productive than chickens reared in environments that do not boost their health. Enriched battery cages increase the potential of egg production in chickens.

In fact, recent estimations reveal that sixty percent of eggs used in products such as mayonnaise, sandwiches and cakes are from chickens reared in battery cages. This reveals that chickens have more tendencies to produce eggs in battery cages compared to other rearing systems. It is more advisable to use battery cages to aid egg production.

Qualitative Feeding for the Chickens

The battery cage system makes feeding for chickens a habitual process which connotes more care for the chickens. This is because the chickens are fed through a long bisected metal or plastic pipe and water is served to them with overhead nipple systems which are set before the chickens. This makes it easy for the poultry farmer to care for the chickens with frequent and adequate food and water.

Chickens are well-catered for in battery cages and this makes them healthier and more productive.

Battery cages make chickens healthy and more productive.

Low Cost of Labour

Battery cage reduces labour by the poultry farmer. This is because the chickens are well-organized and conducted to be in rows with several facilities for their feeding, carriers of their faeces droppings adequately positioned and the eggs rolling to the appropriate places designed in the battery cages. Everything is organized already.

The poultry farmer will simply do the job of pouring food and water in the designated areas, cleaning the faeces that drop and packing the eggs. This reduces stress for the farmer and the need to employ many workers for poultry farms will be minimized. In the long run, poultry farming will be more rewarding for the poultry workers.

Poultry Battery Cages have Feeders that Minimize Feed Spillage

Battery cages make it easy for poultry farmers to care adequately for their chickens. Despite this dire need to take care of the chickens, battery cages help the poultry farmer to minimize feed spillage when feeding chickens.

It is easy to minimize feed spillage with battery cages because they are designed with standard accessories for holding food and water, especially the amount required for feeding the chickens daily. This disallows any form of feed spillage and wastage. Instead of feed spillage, the chickens keep feeding on the required amount of food specifically provided in front of them.

High Capacity for Accommodation

Modern battery cages have high capacity to accommodate a larger number of chickens within a limited space.

That means, battery cages will save you cost in construction and help you maximize the use of whatever space you have available. Many A-type poultry battery cages can accommodate between 60 and 128 chickens or more per unit, depending on the size and model you buy.

Therefore, it is rewarding to invest in poultry farming using poultry battery cages with high capacity for accommodation.

Affordable, Durable and Suitable for all Egg-laying Chicken Types

Poultry battery cages are quite affordable for beginners and experienced poultry farmers who need to give this method of rearing chickens a trial. It costs less and is more effective and efficient.

Besides being affordable, poultry battery cages are durable because they have wire mesh made of galvanized iron which keeps them from rusting and makes them last for a lengthy period of time. It is not just durable but suitable for all egg-laying chicken types such as breeders, layers, point-of-lay and point-of-cage chickens.

It is therefore advisable for poultry farmers to consider using the modern poultry battery cages for sustainable and adequate productivity in chicken rearing.

FURNISHED CAGES

Furnished cages, sometimes called *enriched* or *modified* cages, are cages for egg laying hens which have been designed to overcome some of the welfare concerns of battery cages whilst retaining their economic and husbandry advantages, and also provide some of the welfare advantages of non-cage systems. Many design features of furnished cages have been incorporated because research in animal welfare science has shown them to be of benefit to the hens.

Specifications

Furnished cages must provide at least the following:

- Area of 750 cm² (120 sq in) per hen, of which 600 cm² (90 sq in) is 45 cm (18") high.
- At least 15 cm (6") of perch per hen.
- At least 12 cm (5") of food trough per hen.
- A nest.
- A claw shortening device.
- A littered area for scratching and pecking.

Furnished Cages and Battery Cages

Furnished cages retain several advantages of battery cages in that they:

- Separate the eggs from the hens' feces thereby keeping the eggs clean.
- Protect the hens from predation.
- Automatically collect the eggs thereby preventing egg-eating and floor-laying which both incur additional cost.
- Retain a small group size which reduces injurious pecking behavior.

Furnished cages have welfare benefits additional to battery cages by providing:

- Additional space,

- A nest,

- A perch,

- A claw shortening device,

- A dust bath/litter substrate,

- Easier access for depopulation (because the stipulation of 5" feed space per bird means that the cages are broad but not deep).

Current Designs

There is no clear limit to the size of the furnished cages. Although initial models were not much larger than conventional battery cages, most current designs house 40 to 80 hens although one system houses 115 hens. The depth of furnished cages is often more than the depth of battery cages and as a result, they are often arranged with only one cage row per level, i.e. not connected back-to-back. The more shallow cages can be connected back-to-back. To create space for large groups of hens, some designs of furnished cages are very long. Cage bottoms are made of wire mesh or plastic slats and are sloped so that eggs not laid in the nest box roll onto an egg belt. Feed is provided in feeders outside the cage, although in some designs there may be internal feeders or a combination of the two. Perches in some designs are raised and in others are at floor level.

Welfare Benefits

In a study which compared the welfare benefits of hens in furnished cages, battery cages, free range and barn systems, hens in furnished cages had the lowest faecal corticosterone (a hormone that indicates stress levels), the lowest number of hens that were vent pecked, lowest number of egg shells with calcium spots (an indicator of stress when the egg is temporarily retained by the hen), lowest number of egg shells with blood spots on (usually caused by prolapse), lowest score of skin damage, lowest severity of vent damage caused by vent pecking and lowest plumage soiling. Hens in furnished cages had a similar percentage of hens with recent keel fractures which are usually caused during depopulation (3.6%) compared to hens in barn (1.2%) and free-range systems (1.3%), all of which were considerably lower than in hens from battery cages (24.6%). Furthermore, hens in furnished cages had a smaller percentage of old keel fractures (31.7%) compared to hens in barn (69.1%) and free-range (59.8%) systems but more than hens in battery cages (17.7%). This indicates that furnished cages protect against the keel breaks that are common amongst non-caged hens and also protects against the effects of osteoporosis prevalent in battery cages causing bones to be weak and easily broken during depopulation. In this study, mortality rates were above the breed standards in all systems except the furnished cages.

Welfare Disadvantages

Furnished cages provide more space than battery cages but still prevent some behaviors such as vigorous wing-flapping, flying, nest-building (no materials are provided) and inhibit others (comfort or grooming behaviors) determined partly by the numbers of hens in the cage. The hens are not separated from their feces as completely as hens in battery cages and therefore are at a greater risk of disease, although not as great as the risk to hens in non-cage systems. The small amount of litter that is provided in furnished cages is often distributed quickly or flicked out the cage, possibly resulting in frustration for hens wishing to dustbath and resulting in sham dustbathing. The nest boxes are often occupied by hens using the box for behaviors other than egg-laying (e.g. for sleeping or sham dustbathing) which could lead to frustration in hens wishing to lay an egg.

Production in Furnished Cages

Some studies indicate that production in furnished cages is comparable to that in battery cages. Other studies indicate hens housed in furnished cages have better body-weights and egg production compared to hens in battery cages.

DEBEAKING

An adult bird which has been beak-trimmed as a chick.

Debeaking is the partial removal of the beak of poultry, especially layer hens and turkeys although it may also be performed on quail and ducks. Most commonly, the beak is shortened permanently, although regrowth can occur. The trimmed lower beak is somewhat longer than the upper beak.

Beak trimming is most common in egg-laying strains of chickens. In some countries, such as the United States, turkeys routinely have their beaks trimmed. In the UK, only 10% of turkeys are beak trimmed. Beak trimming is a preventive measure to reduce damage caused by injurious pecking such as cannibalism, feather pecking and vent pecking, and thereby improve livability. Commercial broiler chickens are not routinely beak trimmed as they reach slaughter weight at approximately 6 weeks of age, i.e. before injurious pecking usually begins. However, broiler breeding stock may be trimmed to prevent damage during mating. In some countries, beak trimming is done as a last resort where alternatives are considered not to be possible or appropriate.

Opponents of beak trimming state that the practice reduces problem pecking by minor amounts compared to the trauma, injury, and harm done to the entire flock by beak trimming. Reduction is in single digit percentiles, whereas improvement of conditions especially in layer colonies will cease problematic behavior entirely.

Beak trimming has been banned in Switzerland since 1992 and has been phased out in Germany in 2017.

In close confinement, cannibalism, feather pecking and aggression are common among turkeys, ducks, pheasants, quail, and chickens of many breeds (including both heritage breeds and modern hybrids) kept for eggs. The tendency to cannibalism and feather pecking varies among different strains of chickens, but does not manifest itself consistently. Some flocks of the same breed may be entirely free from cannibalism, while others, under the same management, may have a serious outbreak. Mortalities, mainly due to cannibalism, can be up to 15% in egg laying flocks housed in aviaries, straw yards, and free-range systems.

Because egg laying strains of chickens can be kept in smaller group sizes in caged systems, cannibalism is reduced leading to a lowered trend in mortality as compared to non-cage systems. Cannibalism among flocks is highly variable and when it is not problematic, then mortalities among production systems are similar.

Beak trimming was developed at the Ohio Experiment Station in the 1930s. The original technique was temporary, cutting approximately 6 mm (1/4 inch) off the beak. It was thought that the tip of the beak had no blood supply and presumably no sensation. The procedure was performed by hand with a sharp knife, either when deaths due to cannibalism became excessive, or when the problem was anticipated because of a history of cannibalism in the particular strain of chicken.

Cannibalism is a serious management problem dating back to the periods before intensive housing of poultry became popular. Poultry books written before vertical integration of the poultry industry describe the abnormal pecking of poultry.

Chicks and adult birds' picking at each other until blood shows and then destroying one another by further picking is a source of great loss in many flocks, especially when kept in confinement. The recommendation of the Ohio Experiment Station of cutting back the tip of the upper beak has been found to be effective until the beak grows out again.

Cannibalism has two peaks in the life of a chicken; during the brooding period and at the onset of egg laying. The point-of-lay cannibalism is generally the most damaging and gets most of the attention. The temporary beak trimming developed at the Ohio

Experiment Station assumed that cannibalism was a phase, and that blunting the beak temporarily would be adequate.

Current Methods and Guidelines

In recent years, the aim has been to develop more permanent beak trimming (although repeat trimming may be required), using electrically heated blades in a beak trimming machine, to provide a self-cauterizing cut. There are currently four widely used methods of beak trimming: hot blade, cold blade (including scissors or secateurs), electrical (the Bio-beaker) and infrared. The latter two methods usually remove only the tip of the beak and do not leave an open wound; therefore they may offer improvements in welfare. Other approaches such as the use of lasers, freeze drying and chemical retardation have been investigated but are not in widespread use. The infrared method directs a strong source of heat into the inner tissue of the beak and after a few weeks, the tip of the upper and lower beak dies and drops off making the beak shorter with blunt tips. The Bio-beaker, which uses an electric current to burn a small hole in the upper beak, is the preferred method for trimming the beaks of turkeys. The Farm Animal Welfare Council (FAWC) wrote regarding beak trimming of turkeys that cold cutting was the most accurate method, but that substantial re-growth of the beak occurred; although the Bio-beaker limited beak re-growth, it was less accurate. It was considered that the hot cut was the most distressing procedure for turkeys.

In the UK, beak trimming of layer hens normally occurs at 1-day of age at the same time as the chick is being sexed and vaccinated.

USA's UEP guidelines suggest that in egg laying strains of chickens, the length of the upper beak distal from the nostrils that remains following trimming, should be 2 to 3 mm. In the UK, the Farm Animal Welfare Council stated: "The accepted procedure is to remove not more than one third of the upper and lower beaks or not more than one third of the upper beak only" but went on to recommend: "Where beak trimming is carried out, it should, wherever possible, be restricted to beak tipping; that is the blunting of the beak to remove the sharp point which can be the cause of the most severe damage to other birds."

Costs and Benefits

Costs

The costs of beak trimming relate primarily to welfare concerns. These include acute stress, and acute, possibly chronic, pain following trimming. A bird's ability to consume food is impaired following beak trimming because of the new beak shape and pain. Most studies report reduced body weights and feed intake following beak trimming; however, by sexual maturity or peak egg production, growth rates are usually normal. Weight losses were reduced in chicks that were beak trimmed by infrared compared with chicks trimmed by a hot-blade.

The Pain of Beak Trimming

White Leghorn pullets showing the results of beak trimming.

Non-beak trimmed.   Beak trimmed.

Acute Pain

The beak is a complex, functional organ with an extensive nervous supply including nociceptors that sense pain and noxious stimuli. These would almost certainly be stimulated during beak trimming, indicating strongly that acute pain would be experienced. Behavioral evidence of pain after beak trimming in layer hen chicks has been based on the observed reduction in pecking behavior, reduced activity and social behavior, and increased sleep duration. In Japanese quail, beak-trimming by cauterization caused lower body weights and feed intake in the period just after beak trimming. Beak trimmed Muscovy ducks spent less time engaging in beak-related behaviors (preening, feeding, drinking, exploratory pecking) and more time resting than non-trimmed ducks in the days immediately post-trim. These differences disappeared by 1 week post-trim. At 1 week post-trim the trimmed ducks weighed less than non-trimmed ducks, but this difference disappeared by 2 weeks post-trim. It is, however, unclear if the above changes in behavior arise from pain or from a loss of sensitivity in the beak. Pecking force has been found to decrease after beak trimming in adult hens possibly indicating that hens are protecting a painful area from further stimulation. However, pecking force did not differ between chicks with or without minor beak-trims at 2 to 9 days of age, suggesting that chicks with minor beak-trims do not experience pain from the beak.

Chronic Pain

Severe beak trimming or beak trimming birds at an older age are thought to cause chronic pain. Following beak trimming of older or adult hens, the nociceptors in the beak stump show abnormal patterns of neural discharge, which indicate acute pain. Neuromas, tangled masses of swollen regenerating axon sprouts, are found in the healed stumps of birds beak trimmed at 5 weeks of age or older and in severely beak

trimmed birds. Neuromas have been associated with phantom pain in human ampu-
tees and have therefore been linked to chronic pain in beak trimmed birds. If beak
trimming is severe because of improper procedure or done in older birds, the neuromas
will persist which suggests that beak trimmed older birds experience chronic pain, although this has been debated.

Benefits

The benefits of beak trimming are mainly welfare advantages for birds kept in close confinement, some of which directly relate to increases (or reduced decreases) in production.
These include reduced feather pecking and cannibalism, better feathering (though they
find it hard to preen with shortened beaks, which means they are not cleaning themselves
well), less fearfulness and nervousness, less chronic stress, and decreased mortality.

Alternative

A range of options have been proposed as possible alternatives to beak trimming including modifying the genetics of domesticated poultry to reduce cannibalistic tendencies. For confined housing where light control is possible, lowering light intensity so
that birds cannot see each other as easily reduces antagonistic encounters and aggressive behavior. Enrichment devices, introduced at an early age, such as simple objects
hung in a habitat, can reduce aggressive behavior. Dividing the population into smaller
group sizes reduces cannibalism. Proper body weight management that avoids underweight pullets reduces the probability of underweight pullets with uterine prolapse that
leads to cloacal cannibalism.

CHICK CULLING

Chick culling is the process of killing newly hatched poultry for which the industry has
no use. It occurs in all industrialised egg production whether free range, organic, or battery cage—including that of the UK and US. Because male chickens do not lay eggs and
only those on breeding programmes are required to fertilise eggs, they are considered
redundant to the egg-laying industries and are usually killed shortly after being sexed,
which occurs after they hatch. Many methods of culling do not involve anaesthetics and
include cervical dislocation, asphyxiation by carbon dioxide and maceration using a
high speed grinder. Asphyxiation is the primary method in the United Kingdom, while
maceration is the primary method in the United States. By 2020, US producers expect
to sex the eggs before they hatch, so male eggs can be culled.

Due to modern selective breeding, laying hen strains differ from meat production
strains (broilers). In the United States, males are culled in egg production, since males
"don't lay eggs or grow large enough to become broilers".

Ducklings and goslings are also culled in the production of foie gras. However, because males put on more weight than females in this production system, the females are culled.

Prior to the development of modern broiler meat breeds, most male chickens (cockerels) were slaughtered for meat, whereas females (pullets) would be kept for egg production. However, once the industry bred separate meat and egg-producing hybrids, there was no reason to keep males of the egg-producing hybrid. As a consequence, the males of egg-laying chickens are killed as soon as possible after hatching and sexing to reduce losses incurred by the breeder. Special techniques have been developed to accurately determine the sex of chicks at as young an age as possible.

As of 2018, worldwide around 7 billion day-old male chicks were culled per year in the egg industry.

Ducklings and Goslings are also culled in the production of foie gras. After hatching, the ducklings and goslings are sexed. Males put on more weight than females, so the females are killed, sometimes in an industrial macerator. Up to 40 million female ducks per year may be killed in this way. The remains of female ducklings are later used in cat food, fertilisers and in the pharmaceutical industry.

Methods

Chick grinding machine.

Several methods are used to cull chicks:

- Maceration: The chicks are placed into a large high-speed grinder.

- Cervical dislocation: The neck is broken.

- Electrocution: An electric current is passed through the chick's body until it is dead.

- Suffocation: The chicks are placed in plastic bags.

- Gases or gas mixtures: Carbon dioxide is used to induce unconsciousness and then death.

US Recommended Methods

The American Veterinary Medical Association recommends cervical dislocation, maceration, and asphyxiation by carbon dioxide as the better options. The 2005-2006 American Veterinary Medical Association Executive Board proposed a policy change, which was recommended by the Animal Welfare Committee on disposal of unwanted chicks, poults, and pipped eggs. The policy states "Unwanted chicks, poults, and pipped eggs should be killed by an acceptable humane method, such as use of a commercially designed macerator that results in instantaneous death. Smothering unwanted chicks or poults in bags or containers is not acceptable. Pips, unwanted chicks, or poults should be killed prior to disposal. A pipped egg, or pip, is one where the chick or poult has not been successful in escaping the egg shell during the hatching process."

Alternatives

A Unilever spokesperson has been quoted as saying "We have also committed to providing funding and expertise for research and introduction of alternative methods such as in-ovo gender identification (sexing) of eggs. This new technology offers the potential to eliminate the hatching and culling of male chicks."

FORCED MOLTING

Forced molting typically involves the removal of food and/or water from poultry for an extended period of time to reinvigorate egg-laying.

Forced molting, sometimes known as induced molting, is the practice by some poultry industries of artificially provoking a flock to molt simultaneously, typically by withdrawing food for 7–14 days and sometimes also withdrawing water for an extended period. Forced molting is usually implemented when egg-production is naturally decreasing toward the end of the first egg-laying phase. During the forced molt, the birds cease producing eggs for at least two weeks, which allows the bird's reproductive tracts to regress and rejuvenate. After the molt, the hen's egg production rate usually peaks slightly lower than the previous peak, but egg quality is improved. The purpose of forced molting is therefore to increase egg production, egg quality, and profitability of flocks in their second or subsequent laying phases, by not allowing the hen's body the necessary time to rejuvenate during the natural cycle of feather replenishment.

The practice is controversial. While it is widespread in the US, it is prohibited in the EU.

Commercial hens usually begin laying eggs at 16–20 weeks of age, although production gradually declines soon after from approximately 25 weeks of age. This means that in many countries, by approximately 72 weeks of age, flocks are considered economically unviable and are slaughtered after approximately 12 months of egg production, although chickens will naturally live for 6 or more years. However, in some countries, rather than being slaughtered, the hens are force molted to re-invigorate egg-laying for a second, and sometimes subsequent, laying phase.

Forced molting simulates the natural process where chickens grow a new set of feathers in the Autumn, a process generally accompanied by a sharp reduction or cessation of egg production. Natural molting is stimulated by shortening day lengths combined with stress (of any kind). Before confinement housing with artificial lights was the norm, the Autumn molt caused a seasonal scarcity of eggs and high market prices. Farmers attempted to pamper their flocks to prevent the molt as long as possible, to take advantage of the high prices. Modern controlled-environment confinement housing has the opposite problem; the hens are not normally presented with sufficient stress or cues to go into molt naturally. However, after laying almost daily for nearly a year, their rate of egg production declines, as does the quality of the eggshell and the egg contents. In addition, the hens are overweight.

It is sometimes claimed that forced molting is an artifact of modern intensive farming, but the practice predates the vertical integration of the poultry industry by decades.

Methods

For a complete recovery of the reproductive tract, the hen's body weight must drop by 30 to 35 percent during the forced molt. This is typically achieved by withdrawing the hen's feed for 7–14 days, sometimes up to 28 days. This induces the birds to lose their feathers, cease to lay eggs and lose body-weight. Some programs combine feed withdrawal with a period of water withdrawal. Most programs also restrict the amount of lighting to provide a daylight period that is too short to stimulate egg production, providing a simulated autumn, the natural time of molt and minimum egg production.

Forced molting programs sometimes follow other variations. Some do not eliminate feed altogether, but may induce a molt by providing a low-density diet (e.g. grape pomace, cotton seed meal, alfalfa meal) or dietary manipulation to create an imbalance of a particular nutrients. The most important among these include manipulation of minerals including sodium, calcium, iodine and zinc, with full or partially reduced dietary intakes. These alternative methods of forced molting have not been widely used by the egg industries.

In 2003, more than 75% of all flocks in the US were molted.

Mortality

Some birds die during forced molting and it has been recommended that the flock must be managed so that mortality does not exceed 1.25% over the 1–2 weeks of (nearly complete) feed withdrawal, compared to a 0.5% to 1.0% monthly mortality in a well-managed flock under low-stress conditions. Alternative methods of forced molting which do not use total food withdrawal, e.g. creating a dietary mineral imbalance, generally result in lower mortality rates.

PASTURED POULTRY

Pasture poultry is an intensive method of raising a large number of chickens on a small piece of land. Under this system, birds are housed in mobile shelters with roofs and side walls but no floors; the birds sit on the grass under the shelters. Because chickens do not herd as easily as geese or turkeys, the shelters provide an easy means of corralling the birds and moving them from one spot to another in the pasture. By moving the shelters daily to a fresh patch of grass, the grass is not trampled by the birds and stays in better condition than under traditional range rearing. More chickens can be kept per acre because the grass stays fresher and may remain as a soft bedding even when a large number of birds are raised in a field. When properly managed, 1,000 chickens can be raised in shelters on one acre of pasture compared to 250 birds under free range management.

One of the keys to proper management is to never put the birds out on pasture before six weeks of age; you must wait until the birds are fully feathered and no longer need supplemental heat. In years past it was common to start day-old chicks in range shelters with coal or oil stoves but the pasture poultry shelters used today do not have stoves and birds should not be moved into them until the brooding period is completed. Modern meat-type chickens are especially susceptible to chilling and chilling has caused mortality as high as 80% in some small flocks in Manitoba. Chilling early in life can elevate mortality throughout the life of the flock. To keep the birds comfortable before moving them out to pasture, you also need to provide at least 0.1 m² (1 ft.²) of floor space per bird up to six weeks of age.

Once the chickens are in the pasture shelters, it is recommended that you provide 0.15 m2 (1.5 ft.²) of floor space per small broiler (1.8 kg or 4 lb. live weight) or 0.2 m² (2 ft.²) per roaster (2.7 kg or 6 lb. live weight). Remember that meat-type chickens today reach heavier weights at younger ages than in the past and recommended densities are lower than with old-fashioned breeds. This space requirement is particularly necessary during periods of heat stress. Because you are confining the birds to a shelter with no fans or other reliable ventilation, you must give the birds room to spread out and help stop heat from building up in the shelter. You should be able to remove any solid walls or cloths covering the sides of the shelters on hot days; you want as much breeze as possible to flow over the birds. Remember that the birds are confined to a small area and you must provide them the conditions that they need to be comfortable.

If the shelters are moved frequently and precipitation is adequate, the pasture will provide soft bedding for the birds. Because properly managed pasture is less abrasive, chickens raised on pasture may have a lower incidence of breast blisters. Moving the shelter regularly will reduce the number of droppings that the chickens eat and lower the chances that they will become infected with coccidiosis (the most common type of intestinal parasite in chickens). Since coccidia can survive eight weeks in a field, it is important to never return the birds to the same spot in the pasture in the same year. Other intestinal parasites can survive more than a year in a field and so you should never have chickens on the same field two years in a row.

POULTRY REARING

Basically two systems are commonly followed in poultry rearing:

1. Cage system,

2. Deep litter system.

Cage system: The cage system of rearing birds has been considered as a super intensive system providing floor area of 450-525 sq.cm. (0.6-0.75 sq.feet) per bird. In cage the birds are kept in one, two or three per cage, arranged in single or double or triple rows.

Cage system

Advantages

1. Greater number of birds is reared per unit of area.

2. Facilitates correct maintenance of records.

3. Helps in identifying poor producers and prompt culling.

4. Control of vices of poultry cannibalism and egg eating.

5. It helps in production of clean eggs.

6. Removal of stress factors.

7. Easy control of parasitic disease like coccidiosis and worm infestation.

8. Prompt steps to control feed wastage.

9. The cage method of housing is ideal for the area of moderated climate conditions where the day temperature in summer does not high and temperature does not fall too low.

10. Egg production of caged layer was reported to be more then those kept in deep litter system.

11. Feed efficiency and egg weight were better in caged birds than the laying flock under deep litter system.

Disadvantages

1. Difficulties in ensuring proper ventilation to birds especially in summer season and under very high densed conditions.

2. Incidence of leg problem, cage layer fatigue, fatty liver syndrome, flies and obnoxious gases in the house will be on increases.

3. Hysteriosis of chicks.

Cage fatigue is considered to a physiological derangement of mineral electrolytes imbalance. Leg weakness is common in caged birds.

Cage fatigue.

Fatty live syndrome: It is a problem met with caged layers due to increased deposition of fat in the body resulting in death due to internal hemorrhage. Increasing the protein level and the diet strengthened by the addition of choline, vitamin B12, inositol and vitamin-E may be helpful in reducing the incidence of problem.

Proper ventilation, correction of light-intensity, duration, temperature, ideal environmental conditions, and maintenance of comfort in cages will check the conditions of hysteria of chicken in cages.

Deep litter system: Deep litter system is commonly used in all over the world.

Deep litter system.

Advantages

1. It is an economical.

2. Hygienic, comfortable and safe to birds.

3. Built up litter supplies vitamin B12 and Riboflavin to the birds.

4. Controls diseases and vices.

5. It increases the efficiency of production.

6. Materials such as paddy husks saw dust, dried leaf, chopped straw and groundnut kernels depending upon the availability can be used as litter materials.

Points to be considered while adopting deep litter system:

1. The deep litter system should always kept dry.

2. Only right numbers of birds should be housed.

3. The house should be well ventilated.

4. The litter should be stirred at least once in a week-wet litter if any should be replaced immediately with new dry litter and birds must be fed a balanced ratio.

5. The time starting deep litter system should be in the dry period of the year as it allows sufficient time (At least two months) for bacterial action.

6. Placing of water should be given due attention to keep litter dry.

Confinement Rearing

Size of flock: Larger size units are more economical than smaller ones under commercial conditions. A unit of 2000 layers is usually considered as economical for commercial egg production. In the case of broilers a unit intake of 250 chicks per week is usually considered as viable.

Stock: Procure the best quality chicks. No amount of good management can convert poor quality chicks into good layers or broilers. More profit can be made in a commercial unit by procuring day old pullet chicks. In broiler units, straight-run chicks would give equally good performance.

Number to be procured: In determining the number to be procured, normal losses that might occur due to death and culling have to be allowed. For each 1000 layers to be housed, procure 1100, day-old pullet chicks or 1050 growing pullet chicks or 1000 ready-to-lay pullets. In the case of broilers, the corresponding number would be 250-day-old straight-run chicks for 250 broilers to be marketed at 6-7 weeks of age.

Artificial brooding: Chicks newly hatched out require supplementary heat till they grow feathers. The period of brooding is usually up to 4-5 weeks of age and a little longer in cold season. Artificial brooding can be carried out in deep litter houses or in electrically operated brooder batteries.

Artificial brooding.

Table: Floor space, feeding space and watering space for chicks.

Age weeks	Floor space Sq.ft./Chick	Feeding space inches/chick	Watering space inches/chick
1	0.2	1.5	0.5
2	0.2	2.0	0.7
3	0.3	2.0	0.7
4	0.4	2.5	0.8
5	0.6	2.5	0.8
6	0.8	3.0	1.0
7	0.9	3.0	1.0

On the deep litter, provide 700 cm^2 floor area per chick till 8 weeks of age. In a hover with one m diameter, 250 chicks can be brooded. The hover can be metal or bamboo basket fitted with a heat source. The size and number of the hovers depend on the number of chicks to be brooded. Units of 250 chicks are ideal for efficient management. The hover can be placed at appropriate height from the floor either by hanging it from the roof or by placing it over bricks or stones so that chicks can go in and out easily. Temperature required for brooding is 1 −2 Watt/chick. Use five bulbs of 60 Watts per unit of 250 chicks.

Electricity is the common source of heat used. Electric bulbs of multiple units are preferred over single bulb to cover the wattage. Infra-red bulbs can also be used for brooding. Hover is not necessary when infrared bulbs are used. The number of bulbs to be used depends on the number of chicks to be brooded. The rule of thumb is that one Infra-red bulb of 250 watts for every 250 chicks. Position the bulb 50 cm above litter.

The requirement of chicks for additional warmth decreases as they grow. The warmth as measured by thermometer at 5 cm (2 inches) above the floor level should be checked everyday.

Table: Temperature requirement of chicks during different ages.

Age in weeks	Temperature under hover, at 5 cm above floor (°C)
0-1	35
0-2	32
2-3	29
3-4	26
5-5	23

The distribution of chicks under the hover is a better indication of warmth than the thermometer. If the chicks are active, busy eating and drinking, it indicates that the temperature under the hover is comfortable. Generally one watt per chick appears satisfactory under our climatic conditions.

Litter management: Litter materials such as wood shavings; saw dust, paddy husk, peanut shell, paddy chaff, chopped straw and such other materials that absorb moisture well can be used depending upon the cost and availability. Spread the litter to a depth of 5 cm on the floor before introducing chicks and build it up to a depth of 15 cm by adding litter material, at the rate of about 2 cm per week. This would require approximately 10 kg of litter material/sq.meter. Litter should be raked thoroughly at frequent intervals, say at least twice a week, during the cold and rainy season, once a week during the hot season and the day after deworming. Litter should be kept dry always. During the cold and rainy season and on the area of floor where watering utensils are placed, special attention should be paid daily to check the litter condition. If required, top-dress with fresh litter. It is desirable to use dry lime at the rate of 10 kg per 10 m³ and rake the litter.

Light: Artificial light should be discontinued from the time the chicks no more require additional warmth. Dim light of a 40-watt bulb for every 250 chicks can be provided during the night for broiler chicks.

BROODING MANAGEMENT

Brooding refers to the period immediately after hatch when special care and attention must be given to chicks to ensure their health and survival.

A newly hatched chick has not developed the mechanism to regulate its body temperature therefore, it cannot maintain its body temperature properly for the first few weeks and It is subject to chilling.

When heat is not provided from external sources, the chicks will not take sufficient feeds and water and this leads to the retardation of growth and poor development of internal organs, responsible for digestion,thus the chick will not be able to digest the yolk completely.

Brooding can be classified into natural and artificial brooding:

Natural Brooding

- It is done with the help of broody hens after hatching, up to 3-4 weeks of age.

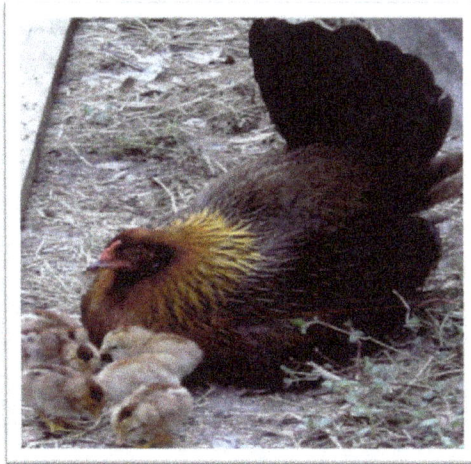

Artificial Brooding

Chicks are reared in the absence of a broody hen. Artificial brooding is mainly aimed at providing the right temperature for the chicks.

Factors to Consider when Brooding

Before Receiving the Chicks

- Brooding house/shed MUST be cleaned.

- Soak the floor preferable with a strong disinfectant.

- Curtains used should be soaked in disinfectant and hanged in the sun to dry.

- Feeders and drinkers should be washed and disinfected.

- Arrange all equipment in the house and spread the litter, prepare the brooder ring and fix the curtains on the open sides to insulate the brooder house.

- Provide foot baths at the entrance with a disinfectant.

How to Prepare a Chick/Brooder Guard

- Use 2.5 m cardboard sheet, aluminum sheet, and coffee wire as brooder guard material to make a circle that uses 8 metres for 100 chicks for 4 weeks.

- Fill the ring with litter material such as wood shaving, straw etc. upto 10 cm thick from the floor.

- Place the heat source at the centre of the brooder ring.

Feed Management during Brooding

- The use of supplemental feeder trays at placement is recommended to help chicks get off to the best start possible.

- Trays should be provided at the rate of 1 per 100 chicks and should be placed between the main feed and drinker lines and adjacent to the brooders.

- Supplemental feeders should be provided for the first 7-10 days.

- The feed trough height should be adjusted so that they rest on the litter for the first 14 days to ensure all birds can easily access feed without having to climb into the feeder.

- Thereafter, feeders should be raised incrementally throughout the growing period so that the lip of the trough or pan is level with the birds back at all times.

Light Management

- Continuous lighting should be provided for the first 48-72 hours post placement.

- It is highly recommended that all flocks are grown under natural light.

Temperature Management

- Ideal brooding temperatures are as measured 5 cm above the litter surface.

- Evening is the best time to observe the chicks and make temperature adjustment.

- Thermometers may not always be available. Therefore, use the behavior of chicks as a guide.

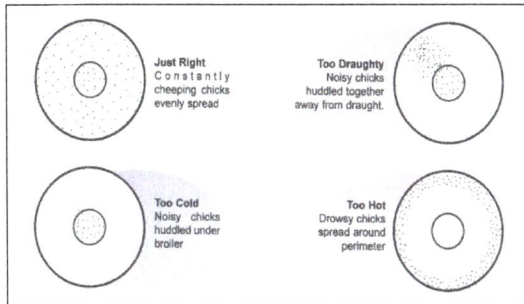

- Adequate floor, feeder and drinking spaces are also important.

- Relative humidity, light and ventilation should be provided for optimum comfort of the chicks.

Source of Heating

- Domestic heaters (jiko) 1 for 100 chicks.

- Infrared lamps (250 watts) 1 for 250 chicks.

- Pancake heater 1 for 1000 chicks.

Main Reasons for Early Chick Mortality

- Poor brooding conditions- high and low brooding temperature.

- Feed poisoning – fungal, toxins, litter poisoning (ingestion of sawdust).

- Injuries- rough handling and pro-longed transportation stress.

- Starvation.

- Humidity.

- Nutrition deficiency.

- Genetic disorder.

- Predators.

Day Old Chicks into the Brooder

- Light the brooder heat source an hour prior to chick arrival so that the ring temperature measure 32 °C.

- Count the chicks and keep records.

- Release the chicks into the brooder ring after dipping their beaks in water.

- Wait for some time to allow the chicks to drink water and keep feed in a chick feeding tray or clean egg tray.

- Do not sprinkle feed on the newspaper as this will get contaminated.

- For the first 3 days watch the chicks at 2-3 hours interval whether they have taken feed and water.

Hay Box Brooder

A hay box is easy to make and is basically a wooden trunk with a top that can be opened or closed. The box is insulated from inside (along the sides) by hay, demarcated by chick mesh wire creating a central warm area where the chicks will sleep.

This is only an overnight box and chicks are taken out during the day. Feed and water are kept out. Provide shelter and make sure the chicks are not exposed to bad weather during the day.

Chickens feeding on grain.

Poultry feed is food for farm poultry, including chickens, ducks, geese and other domestic birds.

Before the twentieth century, poultry were mostly kept on general farms, and foraged for much of their feed, eating insects, grain spilled by cattle and horses, and plants around the farm. This was often supplemented by grain, household scraps, calcium supplements such as oyster shell, and garden waste.

As farming became more specialized, many farms kept flocks too large to be fed in this way, and nutritionally complete poultry feed was developed. Modern feeds for poultry consists largely of grain, protein supplements such as soybean oil meal, mineral supplements, and vitamin supplements. The quantity of feed, and the nutritional requirements of the feed, depend on the weight and age of the poultry, their rate of growth, their rate of egg production, the weather (cold or wet weather causes higher energy expenditure), and the amount of nutrition the poultry obtain from foraging. This results in a wide variety of feed formulations. The substitution of less expensive local ingredients introduces additional variations.

Healthy poultry require a sufficient amount of protein and carbohydrates, along with the necessary vitamins, dietary minerals, and an adequate supply of water. Lactose-fermentation of feed can aid in supplying vitamins and minerals to poultry. Egg laying hens require 4 grams per day of calcium of which 2 grams are used in the egg. Oyster shells are often used as a source of dietary calcium. Certain diets also require the use of *grit*, tiny rocks such as pieces of granite, in the feed. Grit aids in digestion by grinding food as it passes through the gizzard. Grit is not needed if commercial feed is used. Calcium iodate is used as supplement of iodine.

The feed must remain clean and dry; contaminated feed can infect poultry. Damp feed encourages fungal growth. Mycotoxin poisoning, as an example, is "one of the most common and certainly most under-reported causes of toxicoses in poultry". Diseases can be avoided with proper maintenance of the feed and feeder. A *feeder* is the device that supplies the feed to the poultry. For privately raised chickens, or chickens as pets, feed can be delivered through jar, trough or tube feeders. The use of poultry feed can also be supplemented with food found through foraging. In industrial agriculture, machinery is used to automate the feeding process, reducing the cost and increasing the scale of farming. For commercial poultry farming, feed serves as the largest cost of the operation.

Poultry Feed Terms

- *Mash* refers to a nutritionally complete poultry in a ground form. This is the earliest complete poultry ration.

- *Pellets* consist of a mash that has been pelletized; that is, compressed and molded into pellets in a pellet mill. Unlike mash, where the ingredients can separate in shipment and the poultry can pick and choose among the ingredients, the ingredients in a single pellet stay together, and the poultry eat the pellets whole. Pellets are often too large for newly hatched poultry.

- *Crumbles* are pellets that have been sent through rollers to break them into granules. This is often used for chick feed.

- *Scratch grain* (or *scratch feed*) consists of one or more varieties of whole, cracked, or rolled grains. Unlike other feeds, which are fed in troughs, hoppers, or tube feeders, scratch grains are often scattered on the ground. Hence, a large particle size is desired. Because they consist only of grains, scratch grains are not a complete ration, and are used to supplement the balanced ration.

POULTRY HOUSING AND MANAGEMENT

Poultry production has occupied a leading role in the agriculture industry worldwide in recent years. The compound annual growth rate of poultry protein between 2015 and 2025 is estimated to be +2.4%. Asia, South America and Africa characterized by rapid urbanization, poverty and hot climate recorded the highest growth increment in poultry production. The trend of continuous growth of poultry production in those regions is obvious because it remains the fastest route to bridging the protein demand-supply gap.

Extreme weather conditions in the tropical regions of the world have proven generally detrimental to livestock production and is particularly of interest in chicken because of the latter's high sensitivity to temperature change. Just like mammals, the avian species have the ability to regulate their body temperatures by losing or generating heat in response to environmental temperature. If the body temperature of a bird, which normally runs between 39.4 and 40°C, is allowed to increase, the bird will not perform well. Heat stress in poultry production had resulted in under-nutrition, stunted growth, reduction in egg production and size, laying of premature eggs and even death. This problem is further compounded by the high body heat generated by genetically improved laying birds with increased metabolic activity resulting from the high rate of egg production.

Poultry housing design plays a vital role in the determination of the internal climatic conditions of the house for optimum health, growth and productive performance of the birds. Consequently, the type of poultry housing system employed by the proposed poultry farm is a function of the prevailing climatic conditions of the region where the farm is located. While open poultry house system has been adjudged a good method of housing in the tropical countries because of the simplicity of its construction, ease of heat management and minimal management cost, the controlled housing system is the most common in the temperate regions of the world.

Heat Stress in Chicken

Heat Stress is a general problem in the poultry industry, especially in the production of chicken meat and egg. Heat stress is experienced by chicken when the environmental temperature equals or rises above 26.7 °C. At this temperature and beyond the birds begin to pant and can be detrimental to attaining the bird's optimum growth rate, hatching ability, egg size, egg shell quality and egg production. The problem of heat stress can be further compounded in a hot environment when the humidity rises. Heat stress has been reported to have adverse effect on broilers comfort, growth rate, feed conversion, and live weight gain.

In poultry production, the sudden exposure of birds to high temperature short periods is referred to as acute heat stress while exposure for extended periods is referred to as chronic stress. Chronic stress has deleterious effects on birds reared in open-sided houses, which is commonly used in the tropics. It has been reported to have adverse effect on growth and production efficiency, egg quality, meat quality, embryonic development, reproductive performance, immunity and disease incidence's in broilers, laying hens and breeders.

Effects of Internal Climate Conditions on Chicken

It is important we understand the effect of internal climatic conditions of the poultry house on the birds, how the birds respond to them, and their implications on heat management for poultry production. The information will provide guidance on parameters for the open poultry house architectural design that will alleviate heat stress to ensure optimum poultry production in the tropics. The climatic factors of interest include temperature, relative humidity, air composition and velocity, and lighting condition.

Temperature

There is a huge debate on the ideal temperature range required for the various classes and age groups of chicken to attain optimum production. This could be because of other climatic factors such as humidity and wind velocity, which influence temperature change and previous adaptation of chicken to climatic change. Generally, chicken perform under a wide range of temperature regardless of its class (broiler, pullet or breeder) or age. However, exposure of chicken to high temperature has been reported to hinder the performance in chicken production. It could also be further compounded by increased relative humidity for its negative effect on evaporative cooling.

Ketelaars recommended a temperature of 30–32 °C at chicken height for day old chicks. Thereafter, the temperature should be decreased by 3–4 °C till the chicks are 4 weeks old. Daghir reported that a temperature range of 18–22 °C is required for growing broilers. In other reviews done by Holik, it was concluded that birds are comfortable when environmental temperature is within the range of 18–24 °C. However, it should be noted that the optimum performance of chicken is dependent on the market value of the product in relation to feeding cost.

Table: Recommended Temperature Schedule.

Age of chicken (week)	Temperature range (°C)
1	30–32
2	30–26
3	26–23
4	23–20
≥5	20

It is a challenge to maintain the optimum production temperature in the tropics there-fore, it is important that the poultry house designer pay considerable attention to tem-perature change.

Relative Humidity

In a review done by Oloyo, it was reported that internal temperature above 26.7 °C combined with high relative humidity adversely affected the feed efficiency, feathering, pigmentation, and weight gain of chicken. Furthermore, at internal temperature range of 35–37.8 °C the birds' performances were poor regardless of the change in relative humidity. This means higher humidity can improve the performance of the birds at lower temperature. However, humidity must be controlled for it can provide habitat for microorganisms thus, exposing the birds to the threat of disease.

Relative humidity has a strong relationship with temperature change. At the brooding stage, particularly in the earlier weeks the internal relative humidity may be low or too low because of the warming the chicken requires at that age or when the chicks are thirsty or hatched at higher temperature. Soon enough, the internal relative humidity increases because of the water vapor generated by the evaporative cooling act of chick-en to regulate their body temperature as they grow. Consequently, ages 3 weeks and above are very critical periods in chicken production regardless the class of chicken.

In Oloyo, it was reported that laying birds during brooding and after brooding require a relative humidity range of 60–80 and 50–70% respectively for optimum performance.

Air Composition

The decomposition of bird's fecal material produces unpleasant and polluted gases, which include ammonia, carbon dioxide, methane and hydrogen sulphide. These gases are of particular interest because of their adverse effects on the performance of birds, cages, human poultry houses and the environment at large. Consequently, for optimum production for chicken a concentration level of 25 ppm and not more than 2500 ppm for ammonia and carbon dioxide was recommended. It was recommended for good birds' health management that removal of fecal material from the poultry house should be done frequently to reduce the volume of gas emission.

Air Velocity

High internal temperature can be controlled to an extent by varying the air velocity within the poultry house. Also, Air velocity plays an important role in convectional cooling and the regulation of air quality. In hot climatic regions, it is recommended that the ventilation capacity should be at least "5 m^3 per chicken per hour, with inlets amounting to 1.5 cm^2 per m^3 ventilation". Hulzebosch reported that still air velocity (0.1–0.2 m/s) could be maintained if the temperature remains within 25–30 °C. However, Lacy and Czarick, under the same temperature condition reported a better growth rate at 2 and 3 m/s air velocity respectively for broilers.

In the quest to further understand the effect of air velocity on chicken,, factored the ages of chicken within the temperature range 25–30 °C with varying air velocity. The study demonstrated that 6 weeks old broilers benefited from increased air velocity of 2 and 3 m/s than 4 weeks old broilers. This could be because of the high temperature required by younger birds at brooding stage.

Lighting

Lighting at early age in birds have little or no effect on hormonal system, it merely aids birds' activeness including feed intake, growth, and physical and physiological activities. Subsequently, increase in lighting periods and light intensity may cause tiredness, cannibalism, immune responses, leg abnormalities and even death.

The lighting program commonly used is the continuous lighting program of 16 hours light and 8 hours darkness and it has proven successful for overall chicken performance. However, it has been reported that alternating short light and dark period known as intermittent lighting enhances chicken performance. The continuous lighting program with a minimum light intensity of 20 lux is recommended at post-hatch stage (1–7 days old) to help the chick adapt to their environment and aid feeding. Consequently, the light intensity is reduced to about 3–5 lux and intermittent lighting system is introduced for easy control of the birds' activeness for better performance and productivity.

Birds reared under yellow, green, and blue light sources have been reported to have improved body weight compared to those reared under red and orange light sources. Lewis and Morris in a review concluded that the birds reared under blue light show docile trait while those reared under red light were more active and aggressive. In addition, it was noted that the red light improved sexual activities in birds.

Poultry Housing System

The importance of the type of poultry housing system employed for chicken production cannot be over emphasized. It protects the birds from the harsh environmental climatic conditions, which may have adverse effect on the chickens' performance and productivity. In a poultry house, the overall heat generated is the sum of heat generated by the birds, the surrounding environment and biodegradation of fecal material. Thus, the type of housing system to be used is a major determinant factor in the type of management to be adopted in the poultry farm.

Naturally Ventilated Open Housing System

The open poultry housing system has been identified with the tropical region for its simplicity, economic implications and ease of management of heat generation within the building through natural ventilation. However, it is prone to the invasion of insect, rodents, birds and other small predators that can disturb the welfare, productivity and performance of chicken. In the quest to alleviate this problem, dwarf sidewalls are raised to the roof eaves with corrugated wire mesh to keep predators away. Also, gutter filled with insecticides to prevent the invasion of insects are built around the house.

Building Orientation

In order to reduce the exposure of sidewall to direct to direct sun radiation the poultry house should be orientated in the east-west direction. This is very vital, because heat stress in birds can be hastened when they are exposed to direct solar radiation. Deep litter rearing may allow the birds avoid direct sunlight but this may lead to clustering or overcrowding of birds in an area of the house. Consequently, make cooling difficult and in severe cases this leads to stampede and even death.

House Width, Length and Height

The east-west orientation of a poultry house may reduce the benefit of prevailing winds blowing from east or west. Therefore, Daghir recommended that the width of the building should not exceed 12 m to prevent this problem. In addition, the problem of uneven air exchange rate and temperature within the building is eradicated.

Furthermore, the design must factor in the activities and services rendered by poultry farmers and professionals within the building. These activities may include transfer of chicken, feeding, de-pecking, waste management, vaccination, and so on. Therefore, longer pen house could be strenuous to maintain especially when the activities are carried out manually. Doors can be placed at interval of 15–30 m to make for easy circulation and service delivery. Qureshi recommended that for battery cages, it is rather advisable to factor in the number of tiers to be used. Two–tier cage system facilitates easy air exchange within the building whereas, three and four tier cage system can be problematic for air exchange. Therefore, it is recommend that rows of cages should not exceed three with center aisles not less than 1.2 m and a minimum height difference of 1 m from the ceiling.

Roof Slope

A roof slope of 45° was recommended because the angle reduces the heat gain of the roof from the direct solar radiation; maximizes the distance of the bird from the heat accumulated under the roof; quick escape of the heat accumulated under the roof through ridge opening, maximizes air space to improve air exchange rate; and open space above for installation of equipment. On the other hand, the slope in the insulated roof is dependent on the quality of the insulation.

Roof Overhang

Roof overhang can be used to shade the sidewalls of a building from direct and indirect solar radiation. However, the length of the roof overhang is dependent on the height of the sidewalls. Heat gain by the sidewall can be reduced to about 30% by roof overhang shading if properly applied at a roof slope of 45°.

Ridge Opening

Naturally, hot air rises above cooler air due to difference in air density. Introduction of ridge opening can aid ventilation through stack effect in the poultry house. Adequate setback between buildings is required to prevent inadequate airflow and circulation. However, ridge opening has been reported to be ineffective in insulated poultry houses because of temperature uniformity within the house.

Sidewall Openings

The sidewall consists of a dwarf wall built up to the roof eave with a permeable membrane such as a corrugated wire mesh and an adjustable curtain. A minimum height of 0.4 m is recommended to prevent the house from water seepage, direct and indirect solar radiation, pests and predators. The corrugated wire mesh allows easy airflow within and outside the building, while the adjustable curtain is used to control the flow and air velocity. However, the curtain may be transparent or of varying colors to aid its use in managing intermittent lighting scheme.

Building Obstruction

Adequate setback between buildings is required to prevent inadequate air exchange rates in building. Factors such as wind speed, wind direction and topography are major determinants for consideration in defining the optimal house spacing. However, the spacing between buildings can be determined by the expression below:

$$D = 0.4 HL^{0.5}$$

where D, housing spacing (ridge of the closest wall of the next house); H, height of the adjacent building; L, length of the adjacent building.

Vegetation should be kept as minimal as possible and at average height to reduce the nest of wild birds and invasion of rodents and other predators. Also, the branch of trees should be kept at eaves level to prevent obstruction of airflow across the house.

Roof, End-wall and Sidewall Insulation

Farmers in the tropics have successfully used locally sourced materials such as thatched roof and bamboo as roofing materials for the construction of naturally ventilated poultry houses. However, a minimum R-value of $1.25\,m^2\,C/W$ was recommended for ceiling insulation in naturally ventilated poultry house. Environmental temperature higher than 40°C would require a minimum R-value of $2.25\,m^2\,C/W$. The various methods of insulating poultry house ceiling include dropped ceiling, rigid board insulation, spray polyurethane insulation and reflective insulation.

Cooling System

Rooftop sprinklers have proven to be efficient for substantially cooling the roof. However, material of choice in this situation must be able to withstand the constant exposure to water. Evaporative cooling in birds in hot weather can be subdued by using fogging system. With high water pressure it generates mist, which aids cooling in birds. However, the level of humidity within the house must be monitored for it could be detrimental to the health of birds at high temperature. Circulation fan eases heat stress by providing increased air velocity to increase convection cooling. Generally, circulation fans generate air velocity of $0.5\,m/s$ or more and cover an area 15 times its horizontal diameter by five times its vertical diameter. Furthermore, for effective use of circulation fans it should be installed at the center $1-1.5\,m$ above the floor and tilted downward at an angle 5°.

Vegetation

Shrubs and grasses reduce reflective and direct solar radiation by shading and convection cooling. Vegetation should be kept clean and trimmed to keep away predators and pests. The planting of tall trees along the sidewalls can provide a form of canopy to shade the sidewalls from exposure to direct or reflective solar radiation during the hot periods of the day.

Mechanically Ventilated Open Housing System

The limitation of attaining adequate internal environmental conditions required for optimum birds' performance under extreme weather conditions has led to the use of the mechanically ventilated housing system. Also, the mechanically ventilated house provides more control over air exchange, wind velocity and wind direction. Mechanically ventilated system entails the use of either positive or negative pressure system. The negative-pressure system which is the most commonly used in mechanical ventilated house, expels air out of the building by fans through an air inlet system to create low pressure within the house to allow fresh air to rush in through the same air inlet system.

The negative-pressure systems can be achieved through inlet or tunnel ventilation. Inlet ventilation system uniformly distributes exhaust fans and air inlets across the house whereas, tunnel ventilation exhaust fans are located at one end and inlet pipes at the other end. This provides the tunnel ventilation with an advantage of greater air speed in turn creating more positive air exchange.

House Construction

For proper ventilation control, it is required that the house be highly insulated and tightly constructed. However, the sidewall can be equipped with insulated adjustable curtains instead of solid wall for use in the cooler periods of the year or incase of power failure emergency. It is important to note that solid wall have higher insulation value that adjustable curtains.

Air Exchange

High external temperature coupled with the heat generated from the activities within the poultry house increases the temperature of the internal air. An effective mechanical ventilation system has to exchange the air quickly to ensure the internal air temperature maintains not more than 2.8°C difference from the external air temperature. The expression below can be used to calculate the appropriate exhaust fan required for effective ventilation.

$$\text{Building surface heat (watts)} = (A/R) \times (T_o - T_i)$$

where, A, area of the building surface m^2 R, insulation value of the wall material $(m^2 C/W)$; T_o, temperature outside (°C); T_i, temperature inside (°C).

The value of T_o is the hottest external temperature that is excepted of the external environment. However, when calculating heat gain for roof in a house with attic space, the value T_o it is assumed to be 55 °C whereas the T_o value for ceilings with insulation directly below the roof is assumed to be 65 °C. On the other hand, T_i is best assumed as 27 °C to ensure comfort for birds. The value of R will be the overall sum of the insulation value of the wall section.

The total heat produced (sensible and latent) in commercial broiler is 7.9 W/kg while broiler, pullets and broiler breeders is 5.1 W/kg. The heat generated by birds is expressed below.

$$\text{Bird heat (W)} = \text{sensible heat (W/kg)} \times \text{weight of the bird (kg)}$$

where sensible heat, 50% of the total heat produced by birds.

$$\text{Total heat (W)} = \text{building surface heat (W)} + \text{Bird heat (W)}$$

However, the air movement capacity to maintain 2.8 °C between intake and exhaust air is expressed below.

$$\text{AirCapacity}(m^3/h) = \text{total heat(W)} \times 3.4/2.8° C$$

Air Inlet System

There are a number of negative-pressure air inlet pipes used to control the internal climatic condition by controlling the entry location, speed and direction of fresh air. However, the exhaust fan determines how much air enters the house.

Inlet Speed

The pressure difference between the internal and external environment determines the entry speed of fresh air. However, the pressure is a function of the number and sizes of the air inlets. Therefore, the easy manipulation of differential pressure allows for possible control of airflow pattern within the building and of negative-pressure air inlet pipes used to control the internal climatic condition.

Inlet Area

For easy control and distribution of air within the poultry house, the exhaust fan must develop a static pressure of about 12–25 Pa.

Air Inlet Control

Air inlet design should be located strategically as the direction of air depends on external climatic condition, age and class of the chicken. Normally, air inlets should be designed to direct air towards the ceiling at cooler time while another should be directed towards the floor during the hot periods of the year.

Static pressure of about 12–25 Pa was recommended for easy control of the air inlet for a static pressure above or below that range can lead to supply of insufficient air velocity.

Types of Inlet Ventilation System

Cross Ventilation

The exhaust fans are installed on one side while the air inlet pipes are along the other side of the poultry house. It is best suited to narrow poultry houses (less than 10 m) because it leads to difference in environmental conditions in the house with larger width.

Sidewall Inlet Ventilation

The exhaust fans are placed below the air inlet pipes on both sides of the building walls. However, a distance not less than twice the diameter of the fan should be between the exhaust fans and the air inlet pipes. Air movement is directed towards the center, and drawn through the floor by the exhaust fans. It is also best suited for narrow house with not more than 12 m width.

Attic Inlet Ventilation

The exhaust fans are placed on the lower sidewalls while, air inlets are placed in the ceiling. This kind of ventilation requires proper ceiling insulation and it best suitable for hot climate areas. The ventilation method is greatly recommended for rearing laying hen.

Air Movement Inlet Ventilated House

Fresh air enters through the air inlet pipes at a velocity of 3.5-6m/s, however this velocity is quickly dropped to about 1m/s depending on the size and type of the house. Hence, circulation fans are used to boost the air speed to ensure air movement is sufficient in the building.

Tunnel Ventilation System

Tunnel ventilation system is designed to meet the specified air velocity and air exchange rate. However, the required air velocity is dependent on the class of birds in question. Table shows the recommended air speed required for rearing various classes of poultry birds.

Table: Recommended air velocities in tunnel-ventilated houses.

House type	Air speed (m/s)
Broilers	2.5–3
Pullets	1.75–2.25
Broiler breeders	2.25–3
Commercial layer	2.5–3

Tunnel Fan Capacity and Air Velocity

The tunnel fan capacity is determined by the same method used for inlet ventilation system. Unlike the inlet ventilation system where the adequate air velocity is propelled by circulation fan, the required average air velocity within tunnel house is calculated by the expression below.

Air velocity = tunnel fan capacity/(cross- sectional area of the house) $\times$ 3600

where air velocity, m/s; tunnel fan capacity, m^3/h; cross section area, m^2.

However, it is important to note that the cross sectional area of the house adversely affect the air speed within the house. Therefore, it is advisable to design narrow and long house with lower ceilings. Consequently the expression below can be used to design the desired air velocity.

$$\text{Tunnel fan capacity} = \text{desired air velocity}/(\text{cross} - \text{sectional area of the house}) \times 3600$$

where desired air velocity, m/s; tunnel fan capacity, m^3/h ; cross section area, m^2.

In cases where there is land constraints, air deflectors can be installed houses with large cross-sectional area to reduce the cross sectional area within the poultry house. Air deflectors are curtains that extend from the ceiling not more than 2.5–3 m from the ground. Air deflectors have been reported to increase air velocity for a distance approximately 1.2 and 6–9 m upward and downwind of the deflector respectively. However, it is important to ensure that the air deflector exceed 2.5 m from the ground to have it from disrupting the performance of fans and air exchange rate by increasing static pressure.

Air Velocity Distribution

Normally, the air velocity in a tunnel house is assumed uniform across the house. However, it can vary slightly depending on the smoothness of the building surfaces, presence of poultry equipment and other obstructions that deflect air. The difference between the air velocity in the center and the side of the house can vary from 15 to 40%.

Bi-directional Tunnel House

Generally, it is best to install the fans on one end and the inlet in opposite end to ensure the maximum air speed is achieved in the tunnel house. However in cases where the poultry house is over 180 m long and the air velocity required for airflow in one direction exceeds 3.5 m/s it is advisable to apply the bi-directional tunnel house system. The fans are located at end-walls of the building and the tunnel inlet at the center of the house. The air velocity in both direction is reduce to half of the required velocity while retaining the same air exchange rate to ensure the temperature difference between the inlet and the fan remains the same.

Tunnel Fan Placement

The fans can be installed at the end-walls or the sidewalls near the end, and this installation arrangement does not affect the performances of the fans. However, dead spot can be noticed when the fan is installed on the sidewalls as the width of the houses increase.

Tunnel Inlet Opening

In the absence of evaporative cooling pads, it is recommended that the inlet area should be at least 10% greater than the cross sectional area of the house. Meanwhile, the pad used determines inlet size for tunnel house with evaporative cooling pads. It is recommended that inlet opening on the sidewall should be installed as close as possible to the end wall. However, if the house width exceeds 15 m it is advisable to install the inlet openings on the end-wall.

Cool Weather Inlet System for Tunnel Ventilated Houses

It has been recommended that tunnel ventilated system should be used in hot weather because cool weather reduce the air exchange rate. Consequently, it was recommended that a minimum of 60% of the tunnel fan capacity should be controlled by the traditional inlet system before upgrading to tunnel ventilation for easy switch during cooler weathers.

Poultry Exhaust Fans

Exterior and Interior Shutter Fans

It is the simplest type of exhaust fan. Its shutters are used to when the fan is not in use. However, the exterior shutter restricts airflow as air spins off its blades on contact. In the case of interior fan on the other hand, the shutters are on the intake side of the fan thus, lessening the restriction of airflow. It has bigger shutters, which allows for more air movement. Daghir reported that airflow is increased by 5–10% compared to exterior shutter fan.

Discharge Cone Fans

It increases fan performance by 5–10% as it eases the transitioning of drawn towards the fans.

Belt-drive Fans

The fans blades are driven by a simple pulley mechanism. It may be upgraded with an automatic belt tensioner to prevent belt slippage.

Direct-drive Fans

The fan's blades are attached directly to the motor shaft eliminating the use of belt. They are less energy efficient compared to belt-driven fans.

References

- Steiner, Zvonimir; Šperanda, Marcela; Domačinović, Matija; Antunović, Zvonko; Senčić, Đuro (10 July 2006). "Egg quality from free range and cage system of keeping layers". Stockbreeding : Journal of Animal Improvement. 60 (3): 173–179 – via hrcak.srce.hr

- Ani-chik-poultry%20mgt, animal-husbandry: agritech.tnau.ac.in, Retrieved 1 August, 2019

- "United States Standards for Livestock and Meat Marketing Claims". Federal Register. USDA. 30 December 2002. Retrieved 8 January 2013

- 7-key-benefits-of-using-poultry-battery-cages-in-modern-day-chicken-rearing: afrimash.com, Retrieved 2 January, 2019

- Rodenburg, T. B.; Bracke, M. B. M.; Berk, J.; Cooper, J.; Faure, J. M.; Guémené, D.; Guy, G.; Harlander, A.; Jones, T. (December 2005). "Welfare of ducks in European duck husbandry systems". World's Poultry Science Journal. 61 (4): 633–646. Doi:10.1079/WPS200575. ISSN 1743-4777

- Pastured-poultry, poultry, production, livestock, agriculture: gov.mb.ca, Retrieved 3 February, 2019

- "Landmark Ohio Animal Welfare Agreement Reached Among HSUS, Ohioans for Humane Farms, Gov. Strickland, and Leading Livestock Organizations". Humane Society of the United States. 30 June 2010. Retrieved 2 March 2015

- Ani-chik-poultry%20rearing, animal-husbandry: agritech.tnau.ac.in, Retrieved 4 March, 2019

- Gail Damerow (31 January 2012). The Chicken Encyclopedia: An Illustrated Reference. Storey Publishing. Pp. 118–119, 135–136 (for grit). ISBN 978-1-60342-561-2. Retrieved 7 November 2012

- Brooding-management, indigenous-chicken-kienyenj, poultry-chicken, livestock: nafis.go.ke, Retrieved 5 April, 2019

- Horne, P.L.M. Van; Achterbosch, T.J. (2008). "Animal welfare in poultry production systems: impact of EU standards on world trade". World's Poultry Science Journal. Cambridge University Press (CUP). 64 (01): 40–52. Doi:10.1017/s0043933907001705

- Poultry-housing-and-management, online-first: intechopen.com, Retrieved 6 May, 2019

5

Poultry Diseases and their Management

Diseases which afflict poultry such as chicken, turkey and goose are termed as poultry diseases. Some of the common poultry diseases are erysipelas, tick fever and tick paralysis, avian encephalomyelitis, virulent Newcastle disease, fowl pox, haemoproteus, Marek's Disease, etc. Poultry health management aims to prevent and control these diseases. This chapter discusses in detail these poultry diseases and the ways of managing them.

POULTRY DISEASES

Many factors can contribute to diseases in flock. By being aware of their causes and how they spread, you can put practices into place to reduce the risk of disease occurring. Disease can often lead to reduced performance in areas such as breeding, growth rate, feed conversion and egg production. Disease can also affect appearance in show birds and racing ability in pigeon flocks. Although there are many possible causes of disease, it is often a combination of factors that make birds sick.

Infectious Agents

Infectious agents are living organisms that cause disease or illness and can be spread from bird to bird. These include 'germs' (bacteria, viruses, fungi), external parasites (lice and mites) and internal parasites (worms, coccidiosis, blackhead). Infectious agents that cause disease are also referred to as pathogens.

Environmental Conditions

Some environmental conditions can also make birds sick. Unlike infectious agents, the illness is not spread between birds. When the environment affects the health of birds it is usually because the animals are unable to adapt to the conditions. Environmental factors that can cause disease include:

- Poisons.

- Injury.

- Nutritional deficiencies.

- Poor air quality.

- Temperature extremes.

- Physical stress.

- Exposure to disease carrying vermin and insects such as rodents and darkling beetles.

Stress

Severe physical stress can reduce the birds' ability to resist disease. Flocks rely on people to give them:

- Appropriate feed and clean, uncontaminated water.

- Appropriate environmental conditions.

- Shelter.

Without these, birds may suffer stress from fear, malnutrition, dehydration, over-crowding, dirty conditions and extremes in climatic conditions.

Spreading of Infectious Agents

Disease can be spread by new infectious agents entering flock or by the spread of established infectious agents that are already in the flock.

Disease can be spread by:

- People - including through clothing, hands and footwear.

- Domestic and wild birds - through droppings, feathers and discharges from the nose and mouth.

- Contaminated equipment and vehicles.

- Eggs.

- Air.

- Feed and water.

- Animals (e.g. dogs, cats, rodents).

- Insects (e.g. mosquitoes, flies, beetles) - through the transmission of diseases such as fowl pox, tapeworm, Newcastle disease and Salmonella.

In many cases of infection, birds (and other animals) keep illness at bay and do not appear sick. These so-called 'carriers' do not look sick but can spread disease, often without detection. This is often the case with the food safety pathogen Salmonella, where infected birds do not usually show visual signs of infection.

Specific Classes of Disease

Some poultry diseases are more serious due to their ability to spread quickly and their potential impact on commercial poultry industries.

Prohibited matter poultry diseases include:

- Avian influenza (highly pathogenic).
- Duck virus enteritis (duck plague).
- Duck virus hepatitis.
- Infectious bursal disease (hypervirulent and exotic antigenic variant forms).
- Newcastle disease (virulent).

Restricted matter poultry diseases include:

- Avian influenza (low pathogenic).
- Infectious laryngotracheitis virus.
- Newcastle disease (avirulent).
- Salmonella enteritidis infection in poultry.

ERYSIPELAS

Erysipelas is a bacterial disease caused by infection with Erysipelothrix rhusiopathiae. The disease is most often seen as septicemia, but urticarial and endocardial forms exist. E. rhusiopathiae infects a wide range of both avian and mammalian hosts. The disease has been reported in domestic fowl, feral avian species, and captive wild birds and mammals. Infection in reptiles and amphibians has been reported. The organism has also been isolated from the surface slime on fish (without causing disease), which may serve as a source of infection for other species. From an economic standpoint, turkeys are the most important poultry species affected, but serious outbreaks have occurred in chickens, ducks, and geese. Among affected mammals, swine are the most economically important species, but E. rhusiopathiae infection is also a cause of polyarthritis in lambs.

Erysipelas in poultry is seen worldwide and, although considered a sporadic disease, endemic areas exist.

Etiology

E rhusiopathiae is a facultative, anaerobic bacterium. Two additional genomic species have been described: E. tonsillarum, and most recently E. inopinata, but neither is considered pathogenic for poultry. Morphologically, E. tonsillarum and E. inopinata cannot be distinguished from E rhusiopathiae. E. rhusiopathiae stains gram-positive but tends to decolorize, particularly in older cultures. The organism is small, non-acid fast, nonmotile, does not form spores, and produces no known toxins. There is no flagellum, but a capsule has been demonstrated. The cellular morphology of E. rhusiopathiae is variable. The presence of virulence factors such as neuraminidase play a role in bacterial attachment and invasion of host cells. Cells freshly isolated from tissues during acute infection or from smooth colonies are straight or slightly curved small rods that may occur in short chains. Cells from older cultures or rough colonies tend to become filamentous and may be confused with mycelia. The filamentous form occurs more frequently after repeated passages on artificial media.

E. rhusiopathiae has three colony types and grows readily on ordinary culture media containing the blood or sera of various animals. Growth is enhanced by reducing the oxygen content or increasing the carbon dioxide level to 5%–10%. Optimal incubation temperature is 35°–37 °C, and the optimal pH range is 7.4–7.8.

The organism is not readily destroyed by the usual laboratory disinfectants. It is quite resistant to dessication and can survive smoking and pickling processes. It may survive in litter or soil for various lengths of time. Infected carriers may shed the organism, seeding the environment and making disinfection of premises difficult. It is inactivated by a 1:1,000 concentration of bichloride of mercury, 0.5% sodium hydroxide solution, 3.5% liquid cresol, 5% solution of phenol, quaternary ammonium, chlorine, or 0.5% formalin as long as it is not in organic matter.

Although 26 different serotypes of E. rhusiopathiae were described based on an agar gel diffusion test, some of these serotypes have been assigned to E tonsillarum after its confirmation as a separate species. E. rhusiopathiae serotypes 1, 2, and 5 have been most frequently isolated from poultry.

Epidemiology

Erysipelas occurs sporadically in poultry of all ages. It is ubiquitous in nature and found where nitrogenous substances decompose. Turkeys are susceptible regardless of sex or age, although under field conditions it is more common in older birds. Recent evidence indicates there may be a genetically related resistance in turkeys. The incidence in males is reported to be higher, but this is not supported by experimental data. Erysipelas may affect the fertility of males and may contribute to downgrading and processing losses. Infection results from entrance of the organisms through breaks in the skin, through the mucous membranes such as during artificial insemination, by ingestion of contaminated foodstuffs (particularly cannibalism of infected carcasses), and possibly by mechanical transmission via biting insects. The poultry red mite can harbor the organism and may serve as a mechanical vector. Fighting and cannibalism increase losses.

The organism is shed in feces from infected animals and contaminates the soil, in which it may survive for long periods depending on temperature and pH. Poultry, as well as other animals, may be carriers and shed the organism without showing clinical signs of disease. Carriers can shed from feces, urine, saliva, and nasal secretions. Transmission into poultry houses via rodents can occur.

In non vaccinated flocks, morbidity and mortality rates may reach 40%–50%, but mortality is usually <15%. In vaccinated flocks, some birds may be depressed for a short period and recover. Mortality in vaccinated and nonvaccinated poultry is influenced by the virulence of the organism.

No correlation has been shown to exist between the serotype, chemical structure, or biochemical pattern and the manifestation of the septicemic, urticarial, or endocardial forms of erysipelas.

Clinical Findings

Erysipelas is primarily an acute infection that results in sudden death. In an affected flock, a few birds may be depressed but easily aroused; within 24 hr, a few birds will be dead. Just before death, some birds may be very droopy, with an unsteady gait. Chronic clinical disease in a flock is not usual but does occur; birds may have cutaneous lesions and swollen hocks. Turkeys with vegetative endocarditis usually do not have clinical signs and may die suddenly. Erysipelas should be suspected in flocks that have been artificially inseminated 4–5 days before an episode of death without clinical signs. Rainy, cold weather coinciding with the onset of sexual maturity increases risk of clinical disease. Clinical signs in chickens include general weakness, depression, diarrhea, and sudden death. Most sick birds die. In laying hens, egg production may drop markedly. Decreased egg production and conjunctival edema can be seen in organic, cage-free flocks.

Lesions

At necropsy, a generalized darkening of the skin or various sized areas of diffuse darkening is common. The liver and spleen are usually enlarged and friable and may be mottled. Other gross lesions such as peritonitis, pericarditis, petechiation of the heart, catarrhal exudate in the GI tract, and degeneration of fat associated with the thigh and heart may be noted. Vascular damage and fibrin thrombi are common findings on microscopic examination.

Diagnosis

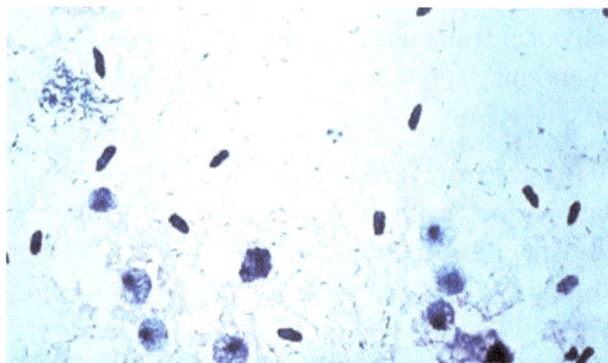

Impression smear of spleen from a turkey infected with
Erysipelothrix rhusiopathiae.

A presumptive diagnosis can be based on an impression smear of the liver or spleen or on a smear of cardiac blood or bone marrow that shows gram-positive, slender, pleomorphic rods. Bone marrow is the tissue of choice in partially decomposed specimens. Isolation and identification of E. rhusiopathiae is necessary for definitive diagnosis. Identification can be made by fluorescent antibody staining and PCR. PCR can distinguish E. rhusiopathiae from E. tonsillarum. A mouse ear scarification model has been described and is particularly helpful for mixed cultures. An indirect ELISA is available, but virtually all poultry are exposed, with antibody levels increasing with age. Caution must be used in attempting reisolation, because the organism produces pinpoint colonies that may be easily overlooked or masked by faster-growing bacteria. Highly selective media are available for reisolation.

Infections with Escherichia coli or Pasteurella multocida, as well as salmonellosis and peracute Newcastle disease, may be confused with the septicemic form of erysipelas. Urticaria and endocarditis may be caused by other miscellaneous bacterial or fungal pathogens. Noninfectious differential diagnoses include poisoning, stampede injuries, or predators.

Treatment, Control and Prevention

The antibiotic of choice is a rapid-acting penicillin such as potassium or sodium penicillin. As soon as a presumptive diagnosis is made, penicillin should be administered IM at 22,000 U/kg body wt, simultaneously with a full dose of erysipelas bacterin. Injectable penicillin is warranted for an acute outbreak but may be an extra-label use. In situations in which it is impractical to handle every bird, administration of penicillin in the drinking water at 395,000 U/L for 4–5 days reduces losses. Sulfonamides and oral oxytetracycline are not effective; broad-spectrum antibiotics, eg, erythromycin, are effective. Antibiotic in feed or water treats only those in the flock that are still eating and drinking normally and may not have dramatic results. Recovered birds have a high degree of resistance. Vaccination with a bacterin helps protect those birds in the flock not yet infected. However, bacterin-derived immunity is not long-lasting and may require two or more injections at intervals of 2–4 weeks. Antibiotic therapy or vaccination does not eliminate the carrier state. Antibiotic resistance to tetracycline has been reported.

Vaccination will control erysipelas. Both inactivated and live vaccines are available for use in turkeys; only vaccines approved for use in turkeys should be used. The use of bacterins in flocks used for meat is useful but labor intensive. For breeders, the bacterin should be given every 4 weeks before onset of egg production. The use of live vaccines administered in the drinking water does not require handling each bird and, therefore, is less stressful. Live vaccines require two doses at intervals of 2–3 weeks. Commercially available swine erysipelas vaccine can be used in layers and is given at up to two times the recommended dose for swine via injection. It does not interfere with egg production. Although the swine vaccine is safe, no challenge studies have been done.

There are no specific husbandry recommendations other than sound management practices for the control of erysipelas in poultry, particularly in endemic areas. After an outbreak, equipment should be thoroughly disinfected and dead birds removed from the premises.

Zoonotic Risks

E. rhusiopathiae can infect people and causes three different syndromes: erysipeloid, a generalized cutaneous form, and a septicemic form with endocarditis. The organism usually enters through cuts in the skin, and those at risk include people who handle infected tissues, such as veterinarians, butchers, and fish handlers. There have been no reports of people becoming infected through the oral route.

TICK FEVER AND TICK PARALYSIS

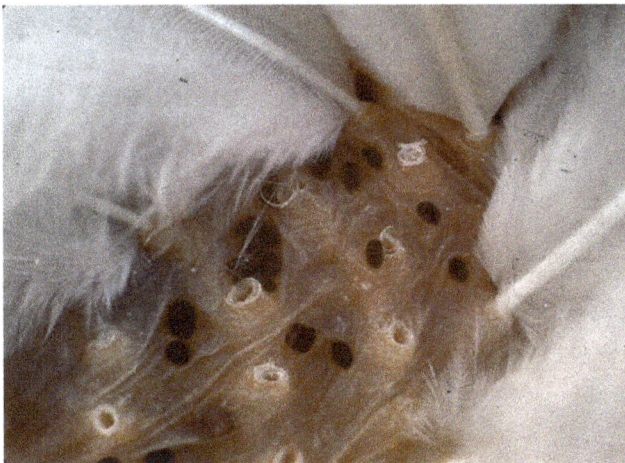

Tick fever is caused by the spirochaete bacterium Borrrelia anserina. The organism is usually transmitted in the bite of an infected fowl tick Argus persicus but it can be spread between birds by ingestion of infective discharges from infected birds. It can be spread mechanically from infected birds by other biting insects but only the tick is its natural host, an infected tick remaining so for well over a year. Fowls that have eaten infected ticks or tick eggs can also contract the disease.

Susceptible Species

Most of the domesticated fowl species are susceptible to tick fever. Pigeons and guinea fowl appear to be resistant. Outbreaks have been seen in some wild bird species overseas.

Incubation Period

The incubation period of the disease varies and depends on the mode of infection. When ticks infect a bird, four to nine days elapse before symptoms develop.

Symptoms

In an acute case, the bird becomes depressed, hangs its head, closes its eyes and passes greenish diarrhoea. The bird feels hot to the touch. Wings and legs may be completely or partially paralysed, the comb is dark and death during convulsions follows, usually within 12 hours.

Symptoms in the chronic form of the disease are less pronounced. Generally, loss of appetite, drowsiness and diarrhoea are present, followed by partial paralysis from which the bird may recover.

The paralysis described in both cases is probably tick paralysis rather than infection with the spirochaete.

Mortality rates in untreated flocks can be high; up to 100% has been recorded.

Diagnosis

The diagnosis can often be based on the clinical picture and the finding of ticks or other heavy infestations of biting parasites.

At post-mortem, the spleen is usually greatly enlarged and has a mottled appearance. The bowel often contains quantities of green mucus with loose contents in the lower bowel.

Laboratory confirmation is made by the use of special stains that reveal the presence of the spirochaete in tissue examined under a microscope.

Treatment

Most antibiotics are effective but penicillin is the drug of choice, one injection often being sufficient to produce a cure in birds not in advanced stages of the disease.

As almost all antibiotics are banned from use in birds producing commercial table eggs, the above advice applies only to non-commercial flocks.

Recovered birds will be immune to reinfection and do not develop a carrier state.

Prevention

There is no longer a tick fever vaccine in Australia so control revolves around adequate control of tick or other external parasites.

The adult parasites do not live on the bird, so when searching for them it is advisable to check all cracks in woodwork by inserting a knife blade or removing timber. Signs of blood indicate that ticks are present. The examination should include roofing timbers, galvanized iron overlaps and loose bark on trees used for perching.

Ticks are extremely hardy and can survive several years without a host. Where sheds are poorly constructed, or birds are not confined to pens, eradication can be difficult.

Control

Parasite numbers can be reduced and controlled but rarely eliminated from housing by spraying insecticide at five-day intervals until thorough inspection fails to detect them. A minimum of three spray applications is necessary (check that the product used is registered for use on poultry and poultry sheds).

Tick Paralysis (Acariasis)

Tick paralysis occurs only from the bite of the tick and is due to a toxin present in the saliva of the tick. This means that tick fever can occur in the absence of tick but tick paralysis only occurs if ticks are present and feeding.

As the bite of one tick may not be sufficient to cause paralysis, a flock into which ticks have recently been introduced, may become immune before the tick numbers increase. Then when a new flock of birds is introduced, they are faced with high tick numbers and become paralysed and die, while the old flock remains unaffected.

Symptoms

The birds usually begin with partial leg paralysis progressing to complete body paralysis and death. In the intermediate phase, they may try and walk using their wings.

Treatment

There is no treatment but if the birds are examined and found to have large numbers of larval or adult ticks on them, they could be dipped in an insecticide to at least stop any more toxin injection.

AVIAN ENCEPHALOMYELITIS

Avian encephalomyelitis (AE) is a viral infection of the central nervous system of poultry, primarily chickens, turkeys, Japanese (coturnix) quail, and pheasants. It is found worldwide and is characterised by ataxia (loss of muscle coordination) and tremors, especially of the head and neck, and a drop in egg production and hatchability in hens. Ducklings, pigeons, and guinea fowl can be infected experimentally. The mortality rate from this disease can be high.

The disease is most common in chickens 1-6 weeks of age. Symptoms usually appear at 7-10 days of age, although they may be present at hatching or delayed for several weeks. Affected chicks may first show a dull expression of the eyes, followed by unsteadiness, sitting on hocks, tremors of the head and neck, paresis (weakness or partial paralysis), and finally total paralysis. Feed and water consumption decreases and birds will lose weight. All stages of the disease can usually be seen in affected flocks. Muscle tremors are best seen after exercising the bird and head tremors are best seen by holding the bird inverted. In adult birds a slight transient drop in egg production may be the only symptom. The disease in turkeys is often milder than in chickens.

What Causes Avian Encephalomyelitis

AE is caused by a picornavirus. Vertical transmission is the most common way the disease is spread but it is also spread by direct contact between susceptible hatchlings and infected birds. Most commercial poultry are exposed in the hatchery when 1 day of age, although further spread occurs later within the flock. The virus present in droppings may survive for more than 4 weeks. Recovered birds are immune and do not spread the virus.

Prevention and Treatment of Avian Encephalomyelitis

There is no treatment for AE. Control of the disease is through prevention. To prevent flocks becoming infected, hatcheries should only accept hatching eggs from immune breeder flocks. Lifetime immunity is acquired through vaccination or recovery from the disease. Breeder pullets should be vaccinated between 9-16 weeks of age. It is also recommended for replacement egg layer pullets to be vaccinated at this age to prevent a temporary drop in egg production.

To minimise the impact of the disease in an infected flock, remove all affected birds and provide good nursing, including fresh food and water, to the remaining birds. Affected birds should be killed and incinerated.

MITES OF POULTRY

The most economically important of the many external parasites of poultry are mites of the families Dermanyssidae (chicken mite, northern fowl mite, and tropical fowl mite) and Trombiculidae (turkey chigger).

Chicken Mite

Dermanyssus gallinae infests chickens, turkeys, pigeons, canaries, and various wild birds worldwide. These bloodsucking mites will also bite people. While rare in modern commercial cage-layer operations, it is found in breeder and small farm flocks. Chicken mites are nocturnal feeders that hide during the day under manure, on roosts, and in cracks and crevices of the chicken house, where they deposit eggs. Populations develop rapidly during the warmer months and more slowly in cold weather; the life cycle may be completed in only 1 week. A house may remain infested for 6 months after birds are removed.

Transmission of the chicken mite, as well as the northern fowl mite and the tropical fowl mite, is by mite dispersion or by contact with infested birds, animals, or inanimate objects. In the integrated poultry industry, mites are dispersed most frequently on inanimate objects such as egg flats, crates, or coops or by personnel going from house to house or farm to farm.

Heavy infestations of either chicken mites or northern fowl mites decrease reproductive potential in males, egg production in females, and weight gain in young birds; they can also cause anemia and death. Chicken mites may be found in the chicken houses during the day, particularly in cracks or where roost poles touch supports, or on birds at night. Their role as vectors of other pathogens in nature needs study, but experimental transmission of Eastern, Western, and Venezuelan equine encephalitis viruses, fowl poxvirus, and the bacteria Salmonella Enteritidis, Pasteurella multocida, Coxiella burnetii, and Borrelia anserina has been demonstrated.

Obtaining mite-free birds and using good sanitation practices are important to prevent a buildup of mite populations. Once poultry have been infested, control may be achieved by spraying or dusting the birds and litter with amitraz, carbaryl, coumaphos, malathion, stirofos, or a pyrethroid compound in areas where the parasites have not developed resistance to these chemicals. Miticide spray treatments must be applied with sufficient force to penetrate the feathers in the vent area. Nicotine sulfate is an effective fumigant for mites but is particularly hazardous. Pyrethrins and piperonyl butoxide are initially active but have poor residual killing power. For control of chicken mites, in addition to treating the birds, the inside of the house and all hiding places for the mite (such as roosts, behind nest boxes, and cracks and crevices) must be treated thoroughly using a high-pressure sprayer. Dimethoate and fenthion may be used as residual house sprays when poultry are not present. Inert dusts such as diatomaceous earth and pure synthetic amorphous silicas can be effective, but application rates need to be high when the humidity is very high. Systemic control with ivermectin (1.8–5.4 mg/kg) or moxidectin (8 mg/kg) is effective for short periods, but the high dosages are expensive, close to toxic levels, and require repeated use.

Common Chigger

The common chigger, Trombicula alfreddugesi, and other chigger species (harvest mites, red bugs) infest birds as well as people and other mammals, feeding on partially digested skin cells and lymph. Heavily parasitized birds become droopy, refuse to eat, and may die from starvation and exhaustion. Larvae may be found either singly or in clusters on the ventral portion of the birds. Control on the range is aided by keeping the grass cut short and dusting with sulfur, carbaryl, or malathion.

Depluming Mite

The depluming mite, Neocnemidocoptes gallinae, is found worldwide and burrows into the epidermis at the base of feather shafts, causing intense irritation and feather pulling and loss in chickens, pheasants, pigeons, and geese in spring and summer. Hyperkeratosis, skin lesions, and digit necrosis can result from the burrowing. Affected birds should be isolated and treated with ivermectin, malathion, or sevin dust.

Feather Mite

Most feather mites belong to the families Analgidae, Pterolichidae, and Proctophyllodidae. Surface feather mites feed mainly on feather oils, debris, fungi, and skin scales. More than 25 species, including Megninia cubitalis, M. ginglymura, and Pterolichus obtusus, are found on domestic poultry, but they are rare on modern poultry ranches. Quill mites (Syringophilidae and Gaudoglyphidae) live in quills and feed on quill tissue or fluids obtained by piercing the calamus wall. Syringophilus bipectinatus is found in chicken and turkey feather quills worldwide, and Columbiphilus polonica, Dermoglyphus elongatus, and Gaudoglyphus minor live in chicken quills in Europe. Feather mites do little economic damage but may reduce egg production via malnutrition, feather loss, and dermatitis. Affected birds should be dusted with pyrethrin or carbaryl powder, or oral or topical ivermectin can be applied.

Northern Fowl Mite

The northern fowl mite, Ornithonyssus sylviarum, is the most important parasite of caged layers and breeding chickens in the USA and is a serious pest of chickens throughout the temperate zone of other countries. On turkeys, it is second in importance only to the turkey chigger in areas where the turkey chigger is found. It has been reported from many species of birds and from rats, mice, and people; however, fertile populations are reported only on birds. Northern fowl mites are obligate bloodsucking parasites that normally spend their entire life cycle (~1 week) on the host. Off the host, mites may live as long as 2 months, depending on temperature and relative humidity. Northern fowl mites are found on eggs or by parting feathers in the vent area, which may have thick, crusty skin, severe scabbing, and soiled feathers.

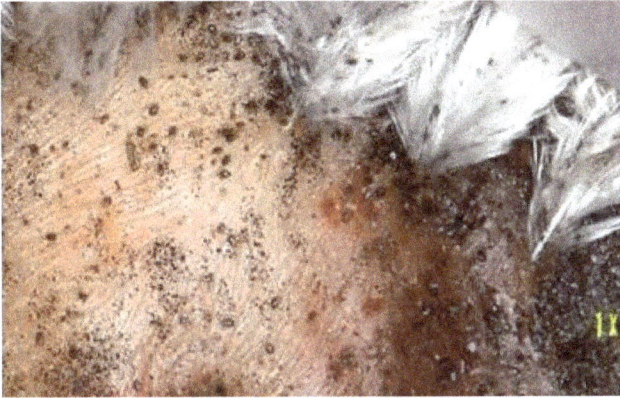

Ornithonyssus sylviarum mites, hen.

Western equine encephalomyelitis, St. Louis encephalitis, and Newcastle disease viruses, as well as fowlpox virus, have been isolated from these mites. However, the mites are not significant vectors of these viruses.

Scaly Leg Mite

The scaly leg mite, Knemidocoptes mutans, is a small, spherical, sarcoptic mite that usually tunnels into the tissue under the scales of the legs. It is rare in modern poultry facilities. When found, it is usually on older birds on which the irritation and exudation cause the legs to become thickened, encrusted, and unsightly. Feet and leg scales become raised, resulting in lameness. Birds stop feeding, and death can result after several months. This mite may occasionally attack the comb and wattles. The entire life cycle is in the skin; transmission is by contact. Infections can be latent for long periods until stress triggers a mite population increase.

For control, affected birds should be culled or isolated, and houses cleaned and sprayed frequently as recommended for the chicken mite. Individual birds should be treated with oral or topical ivermectin or moxidectin (0.2 mg/kg), 10% sulphur solution, or 0.5% sodium fluoride.

Cyst Mite

Laminosioptes cysticola, the fowl cyst mite, is a small cosmopolitan parasite of chickens, turkeys, and pigeons that is most often diagnosed by observing white to yellowish caseocalcareous nodules ~1–3 mm in diameter in the subcutis, muscle, lungs, and abdominal viscera. Careful examination of the skin and subcutis of birds under a dissecting microscope frequently reveals the mites. Destroying the bird has been the best control for this parasite, but ivermectin may be effective.

Tropical Fowl Mite

The tropical fowl mite, Ornithonyssus bursa, is distributed throughout the warmer regions of the world and has been reported in Hawaii, Texas, Florida, Illinois, Indiana, Maryland, and New York. It closely resembles the northern fowl mite in its biology and habits but lays a greater proportion of its eggs in the nest. Hosts include chickens, turkeys, ducks, pigeons, sparrows, starlings, mynah birds, and people. Western equine encephalomyelitis virus has been recovered from this mite, but there is no evidence it transmits the virus.

Turkey Chigger

The larvae of Neoschongastia americana, the turkey chigger, are parasitic on numerous birds. Across the southern USA, they are the major pest of turkeys ranged on heavy clay soils in the summer. The chiggers feed in groups of as many as 100 mites per lesion for 8–15 days. Turkeys may have 25–30 lesions each. One lesion, 3 mm in diameter, may cause significant downgrading at market time. To prevent downgrading, turkeys must be protected for at least 4 weeks before marketing.

Sprays or dusts of carbaryl, malathion, or chlorpyrifos on turkey ranges control chiggers. A preventive measure now used in many turkey-growing areas includes a shift from range to confinement rearing, or use of sheds to provide shade.

VIRULENT NEWCASTLE DISEASE

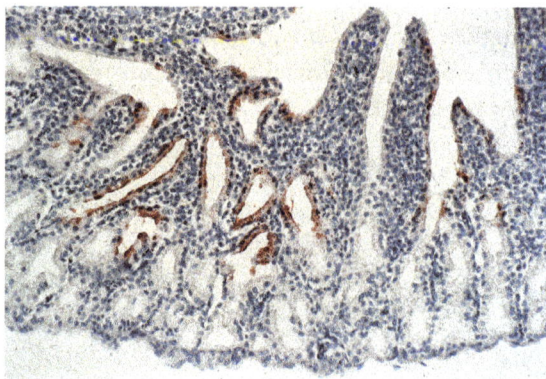

Avian avulavirus 1 (stained in brown) in the conjunctiva of a chicken.

Virulent Newcastle disease (VND), formerly exotic Newcastle disease, is a contagious viral avian disease affecting many domestic and wild bird species; it is transmissible to humans. Its effects are most notable in domestic poultry due to their high susceptibility and the potential for severe impacts of an epizootic on the poultry industries. It is endemic to many countries.

Exposure of humans to infected birds (for example in poultry processing plants) can cause mild conjunctivitis and influenza-like symptoms, but the Newcastle disease virus (NDV) otherwise poses no hazard to human health. No treatment for NDV is known, but the use of prophylactic vaccines and sanitary measures reduces the likelihood of outbreaks.

Newcastle disease was first identified in Java, Indonesia, in 1926, and in 1927, in Newcastle-upon-Tyne, England (whence it got its name). However, it may have been prevalent as early as 1898, when a disease wiped out all the domestic fowl in northwest Scotland.

The policy of slaughter ceased in England and Wales on 31 March 1963, except for the peracute form of Newcastle disease and for fowl plague. In Scotland the slaughter policy continued for all types of fowl pest.

Interest in the use of NDV as an anticancer agent has arisen from the ability of NDV to selectively kill human tumour cells with limited toxicity to normal cells.

Since May 2018, California Department of Food and Agriculture staff and the Department of Agriculture have been working on eliminating VND in South California and more than 400 birds have been confirmed to have VND. On February 27, 2019, California State Veterinarian, Dr. Annette Jones, increased the quarantine area in Southern California and on March 15, 2019 and April 5, 2019, cases of VND in Northern California and Arizona respectively.

Causal Agent

The causal agent, Newcastle disease virus (NDV), is a variant of avian paramyxovirus 1 (APMV-1), a negative-sense, single-stranded RNA virus. NDV/APMV-1 belong to the genus *Avulavirus* in the family *Paramyxoviridae*. Transmission occurs by exposure to faecal and other excretions from infected birds, and through contact with contaminated food, water, equipment, and clothing.

Strains

NDV strains can be categorised as velogenic (highly virulent), mesogenic (intermediate virulence), or lentogenic (nonvirulent). Velogenic strains produce severe nervous and respiratory signs, spread rapidly, and cause up to 90% mortality. Mesogenic strains cause coughing, affect egg quality and production, and result in up to 10% mortality. Lentogenic strains produce mild signs with negligible mortality.

Use as an Anticancer Agent

In 1999, promising results were reported using an attenuated strain of the Newcastle virus, code named MTH-68, in cancer patients by researchers who had isolated the strain in 1968. It appears the virus preferentially targets and replicates in certain types of tumor cells, leaving normal cells almost unaffected. In 2006, researchers from the Hebrew University also succeeded in isolating a variant of the NDV, code named NDV-HUJ, which showed promising results in 14 glioblastoma multiforme patients. In 2011, researchers at the Memorial Sloan–Kettering Cancer Center and the Icahn School of Medicine at Mount Sinai found that NDV modified with the viral protein NS1 had enhanced replication in cancer cell lines overexpressing the antiapoptotic factor Bcl-xL. The researchers suggested in cells that resist the normal inducement of apoptosis when infected will give NDV more time to incubate in cell and spread. Many cancer cells will overexpress antiapoptotic factors as part of tumor development. This mechanism of delaying apoptosis in abnormal cells gives NDV the specificity it needs to be an efficient cancer-fighting oncolytic virus.

Though the oncolytic effect of NDV was documented already in the 1950s, the main advances of viruses in cancer therapy came with the advent of reverse genetics technologies. With these new possibilities, studies of modified NDV strains with enhanced cancer-treatment properties have been put on the agenda. A study demonstrated the engineered Hitcher B1 NDV/F3aa strain could be modified to express a highly fusogenic F-protein in combination with immunostimulatory molecules such as IFN-gamma, interleukin 2, or tumor necrosis factor alpha. Promising results were discovered with proteins associated to the adaptive immune system, which paved the way for possibilities to use NDV to create a tumor-associated antigen. Another study showed how NDV/F3aa could be modified to express NS1, an influenza virus protein with capability to modulate with the innate immune response, for example, by suppressing the induction of the cellular interferons.

NDV Pros and Cons in Cancer Therapy

NDV possesses many unique anticancer properties and thereby provides an excellent base in virotherapy research. NDV has selectivity on oncogenic cells, where it replicates without, or in a less pronounced way, harming normal cells. It binds, fuses into and replicates within the infected cells' cytoplasm independent of cell proliferation. One of the main issues using NDV treatment is the host/patient immune response against the virus itself, which prior to the time of the reverse genetics technology, decreased the applicability of NDV as a cancer treatment.

NDV-induced Mechanisms Leading to Tumor Cell Death

The precise way in which the presence of NDV induces tumor cell death remains to be clarified and may show variation regarding the strains of NDV used and which type of cancer is targeted. NDV triggers apoptosis in a wide range of cancer cell types via the mitochondrial/intrinsic pathway, through loss of membrane potential and thereby inducing release of cytochrome c in the tumor cell. The results also indicate the extrinsic pathway is activated by TNF-related, apoptosis-inducing ligand-induced, NDV-mediated apoptosis in a late stage. Another study found a hyperfusogenic NDV/F3aa (L289A) with refined abilities to fuse into somatic cells. NDV has aggregating properties causing syncytia formations of tumor cells, which, apart from amplifying immune-based cell killing, also results in necrosis of cells. This pathway was believed to lead to a considerable boost of immune activation and potentially an antitumor response, which was supported by observations of a significant accumulation of NK-cells and neutrophils following the infusion of NDV/F3aa(L289A) in hepatocellular carcinoma cells. In addition, an increase of CD4+ and CD8+ T-cells occurs within the tumor cells when inducing NDV/F3aa recombined with the cytokine interleukin-2 (IL-2). An NDV/F3aa-IL-2 strain induced the immune system, giving a cytotoxic effect on the tumor cells. A 15-year study on patients with malignant melanoma showed increased numbers of oligoclonal CD8+ T-cells in the blood, suggesting vaccination with NDV oncolysates was associated with prolonged survival among the patients, and CD8+ T-cells played an important role.

Transmission

NDV is spread primarily through direct contact between healthy birds and the bodily discharges of infected birds. The disease is transmitted through infected birds' droppings and secretions from the nose, mouth, and eyes. NDV spreads rapidly among birds kept in confinement, such as commercially raised chickens.

High concentrations of the NDV are found in birds' bodily discharges; therefore, the disease can be spread easily by mechanical means. Virus-bearing material can be picked up on shoes and clothing and carried from an infected flock to a healthy one.

NDV can survive for several weeks in a warm and humid environment on birds' feathers, manure, and other materials. It can survive indefinitely in frozen material. However, the virus is destroyed rapidly by dehydration and by the ultraviolet rays in sunlight. Smuggled pet birds, especially Amazon parrots from Latin America, pose a great risk of introducing NDV into the US. Amazon parrots are carriers of the disease, but do not show symptoms, and are capable of shedding NDV for more than 400 days.

Clinical Findings

Clinical Signs

Signs of infection with NDV vary greatly depending on factors such as the strain of virus and the health, age and species of the host.

The incubation period for the disease ranges from 4 to 6 days. An infected bird may exhibit several signs, including respiratory signs (gasping, coughing), nervous signs (depression, inappetence, muscular tremors, drooping wings, twisting of head and neck, circling, complete paralysis), swelling of the tissues around the eyes and neck, greenish, watery diarrhea, misshapen, rough- or thin-shelled eggs and reduced egg production.

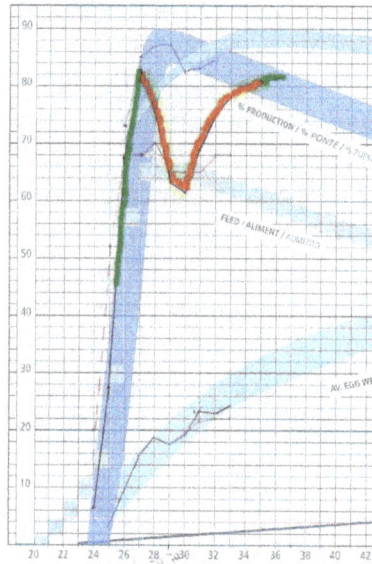

Egg drop after a (otherwise asymptomatic) Newcastle disease
infection in a duly vaccinated broiler parent flock.

In acute cases, the death is very sudden, and, in the beginning of the outbreak, the remaining birds do not seem to be sick. In flocks with good immunity, however, the signs
(respiratory and digestive) are mild and progressive, and are followed after 7 days by
nervous symptoms, especially twisted heads.

Torticollis in a mallard.

Same symptom in a broiler.

PM lesions on proventriculus,
gizzard and duodenum.

Postmortem Lesions

Petechiae in the proventriculus and on the submucosae of the gizzard are typical; also,
severe enteritis of the duodenum occurs. The lesions are scarce in hyperacute cases
(first day of outbreak).

Diagnosis

Immunological Tests

Enzyme-linked immunosorbent assay, polymerase chain reaction, and sequence technology tests have been developed.

Virus Isolation

Samples

For routine isolation of NDV from chickens, turkeys, and other birds, samples are obtained by swabbing the trachea and the cloaca. Cotton swabs can be used. The virus can also be isolated from the lungs, brain, spleen, liver, and kidneys.

Handling

Prior to shipping, samples should be stored at 4 °C (refrigerator). Samples must be shipped in a padded envelope or box. Samples may be sent by regular mail, but overnight is recommended.

Prevention

Any animals showing symptoms of Newcastle disease should be isolated immediately. New birds should also be vaccinated before being introduced to a flock. An inactivated viral vaccine is available, as well as various combination vaccines. A thermotolerant vaccine is available for controlling Newcastle disease in underdeveloped countries.

FOWLPOX

Fowlpox is the worldwide disease of poultry caused by viruses of the family *Poxviridae* and the genus *Avipoxvirus*. The viruses causing fowlpox are distinct from one another but antigenically similar, possible hosts including chickens, turkeys, quail, canaries, pigeons, and many other species of birds. There are two forms of the disease. The first is spread by biting insects (especially mosquitoes) and wound contamination and causes lesions on the comb, wattles, and beak. Birds affected by this form usually recover within a few weeks. The second form is spread by inhalation of the virus and causes a diphtheritic membrane to form in the mouth, pharynx, larynx, and sometimes the trachea. The prognosis for this form is poor.

Fowlpox in Chickens

Fowlpox is a common disease in backyard chickens that have not been vaccinated. Most birds survive the infections, although very young or weak birds may be lost. The lesions initially looks like a whitish blister and appear on the comb, wattles and other skin areas. In rare cases lesions can be found on the body, legs and even sometimes the softer parts of the beak. The blisters develop into a dark scab and take about three weeks to heal and drop off. Fowlpox lesions, when in the infected birds mouth and throat can cause difficulty breathing, even death. Scarring may result and consequently exhibition poultry breeders prefer to vaccinate and avoid this disease. Management of the mosquito population can help reduce outbreaks of fowlpox.

Treatment

Vaccines are available for fowlpox. Chicken are usually vaccinated with *pigeonpox virus*. This vaccine is usually given to chickens when between the age of 8-14 weeks of age, via the wing web method of injection. When a bird is given the vaccine they are exposed to a mild version of the active virus, so they should be completely healthy to prevent severe illness. Turkeys are also routinely vaccinated. Once a bird is infected there are no treatments, just preventative measures including the vaccine and mosquito management.

HAEMOPROTEUS

Haemoproteus spp is the most common blood parasite in birds, especially nondomestic birds. More than 120 species have been reported. Haemoproteus spp are found in free-living ducks, quail, and turkeys but are rare to absent in commercial flocks, probably because of limited vector exposure or very specific feeding habits of Culicoides spp and hippoboscid flies, the invertebrate vectors. Haemoproteus is considered nonpathogenic in most avian species. Birds are usually asymptomatic; however, Haemoproteus infection may be more often significant than previously thought, based on increasing reports documenting decreased host fitness, nestling mortalities, fledging success, and delayed recovery in infected birds versus uninfected birds. Anemia, anorexia, weight loss, and depression have been reported. Rarely fatal disease occurs. Clinical disease is usually attributed to anemia, presence of megaloschizonts in the musculature, or host-cell destruction.

In poultry, infection with H lophortyx in bobwhite quail caused clinical disease and increased mortality with as much as 20% flock loss. Quail 5–10 weeks old were usually affected, and clinical signs included reluctance to move, prostration, and death. Gross lesions included congested spleen and liver and hemorrhagic streaks in the muscles. Histology revealed myositis with megaloschizonts in the musculature, especially of the legs and back, and hemosiderin accumulation in the splenic macrophages. Experimental infection in turkeys with H meleagridis resulted in lameness, diarrhea, anorexia, and depression. Histologic lesions were associated mainly with megaloschizont development in the musculature. Infection in racing pigeons (called pigeon malaria) is commonly asymptomatic but often blamed for poor performances that are due to other diseases or inadequate housing and management. Clinical disease in Columbiformes is rare.

Haemoproteus, Goose

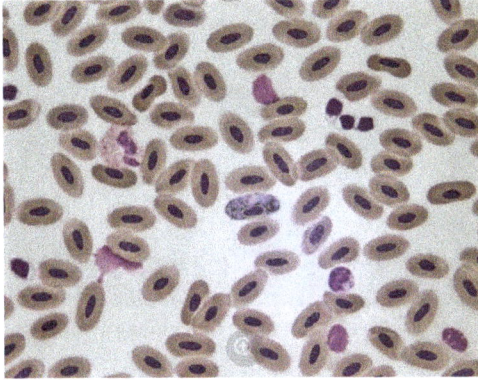

Haemoproteus, goose.

Diagnosis is made by examination of stained blood smears and observation of large, pigmented gametocytes in mature RBCs that partially or occasionally completely encircle the nucleus without displacing it. Merozoites are not seen in the peripheral blood. Schizogony within the endothelial cells of the lung, liver, and spleen may be seen histologically. PCR tests for Haemoproteus have been developed. Little is known about effective treatment. Antimalarial drugs reduce the parasitemia but do not eliminate the parasite. Chloroquine, primaquine, quinacrine, and buparvaquone have been used in pigeons. Combinations of chloroquine and primaquine or chloroquine and mefloquine have been used to treat owls. Treatment is not recommended in asymptomatic birds. Measures to control invertebrate vectors, such as screening of aviaries, help prevent transmission and heavy infections.

MAREK'S DISEASE

Marek's disease is a highly contagious viral neoplastic disease in chickens. It is named after József Marek, a Hungarian veterinarian. Marek's disease is caused by an alphaherpesvirus known as 'Marek's disease virus' (MDV) or *Gallid alphaherpesvirus 2* (GaHV-2). The disease is characterized by the presence of T cell lymphoma as well as infiltration of nerves and organs by lymphocytes. Viruses *related* to MDV appear to be benign and can be used as vaccine strains to prevent Marek's disease. For example, the related Herpesvirus of Turkeys (HVT), causes no apparent disease in turkeys and continues to be used as a vaccine strain for prevention of Marek's disease. Birds infected with GaHV-2 can be carriers and shedders of the virus for life. Newborn chicks are protected by maternal antibodies for a few weeks. After infection, microscopic lesions are present after one to two weeks, and gross lesions are present after three to four weeks. The virus is spread in dander from feather follicles and transmitted by inhalation.

Syndromes

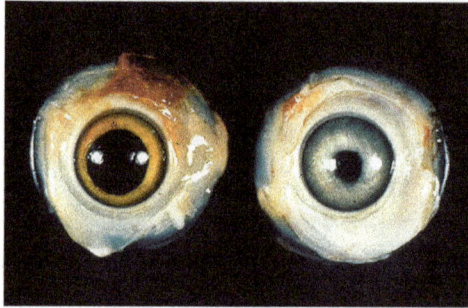

Left—normal chicken eye. Right—Eye of a chicken with Marek's disease.

Six syndromes are known to occur after infection with Marek's disease. These syndromes may overlap.

- Classical Marek's disease or neurolymphomatosis causes asymmetric paralysis of one or more limbs. With vagus nerve involvement, difficulty breathing or dilation of the crop may occur. Besides lesions in the peripheral nerves, there are frequently lymphomatous infiltration/tumours in the skin, skeletal muscle, visceral organs. Organs that are commonly affected include the ovary, spleen, liver, kidneys, lungs, heart, proventriculus and adrenals.

- Acute Marek's disease is an epidemic in a previously uninfected or unvaccinated flock, causing depression, paralysis, and death in a large number of birds (up to 80%). The age of onset is much earlier than the classic form; birds are four to eight weeks old when affected. Infiltration into multiple organs/tissue is observed.

- Ocular lymphomatosis causes lymphocyte infiltration of the iris (making the iris turn grey), unequal size of the pupils, and blindness.

- Cutaneous Marek's disease causes round, firm lesions at the feather follicles.

- Atherosclerosis is induced in experimentally infected chickens.

- Immunosuppression – Impairment of the T-lymphocytes prevents competent immunological response against pathogenic challenge and the affected birds become more susceptible to disease conditions such as coccidiosis and *Escherichia coli* infection. Furthermore, without stimulation by cell-mediated immunity, the humoral immunity conferred by the B-cell lines from the Bursa of Fabricius also shuts down, thus resulting in birds that are totally immunocompromised.

Diagnosis

Diagnosis of lymphoid tumors in poultry is complicated due to multiple etiological agents capable of causing very similar tumors. It is not uncommon that more than one avian tumor virus can be present in a chicken, thus one must consider both the diagnosis of the disease/tumors (pathological diagnosis) and of the virus (etiological diagnosis). A step-wise process has been proposed for diagnosis of Marek's disease which includes (1) history, epidemiology, clinical observations and gross necropsy, (2) characteristics of the tumor cell, and (3) virological characteristics.

The demonstration of peripheral nerve enlargement along with suggestive clinical signs in a bird that is around three to four months old (with or without visceral tumors) is highly suggestive of Marek's disease. Histological examination of nerves reveals infiltration of pleomorphic neoplastic and inflammatory lymphocytes. Peripheral neuropathy should also be considered as a principal rule-out in young chickens with paralysis and nerve enlargement without visceral tumors, especially in nerves with interneuronal edema and infiltration of plasma cells.

The presence of nodules on the internal organs may also suggest Marek's disease, but further testing is required for confirmation. This is done through histological demonstration of lymphomatous infiltration into the affected tissue. A range of leukocytes can be involved, including lymphocytic cell lines such as large lymphocyte, lymphoblast, primitive reticular cells, and occasional plasma cells, as well as macrophage and plasma cells. The T cells are involved in the malignancy, showing neoplastic changes with evidence of mitosis. The lymphomatous infiltrates need to be differentiated from other conditions that affect poultry including lymphoid leukosis and reticuloendotheliosis, as well as an inflammatory event associated with hyperplastic changes of the affected tissue.

Key clinical signs as well as gross and microscopic features that are most useful for differentiating Marek's disease from lymphoid leukosis and reticuloendotheliosis include (1) Age: MD can affect birds at any age, including <16 weeks of age; (2) Clinical signs: Frequent wing and leg paralysis; (3) Incidence: >5% in unvaccinated flocks; (4) Potential nerve enlargement; (5) Interfollicular tumors in the bursa of Fabricius; (6) CNS involvement; (7) Lymphoid proliferation in skin and feather follicles; (8) Pleomorphic lymphoid cells in nerves and tumors; and (9) T-cell lymphomas.

In addition to gross pathology and histology, other advanced procedures used for a definitive diagnosis of Marek's disease include immunohistochemistry to identify cell type and virus-specific antigens, standard and quantitative PCR for identification of the virus, virus isolation to confirm infections, and serology to confirm/exclude infections.

PCR blood testing can also detect Marek's Disease, and proper testing can differentiate between a vaccinated bird with antibodies and a true positive for Marek's Disease.

Marek's Disease is not treatable, however supportive care can help.

It is recommended that all flocks positive for Marek's Disease remain closed, with no bird being introduced or leaving the flock. Strict bio security and proper cleaning is essential, using products like Activated Oxine or Virkon S and reducing dander buildup in the environment. Proper diet, regular deworming and vitamin supplements can also help keep infected flocks healthier. Reducing stress is also a key component, as stress will often bring about illness in birds infected with Marek's Disease.

Prevention

Vaccination is the only known method to prevent the development of tumors when chickens are infected with the virus. However, administration of vaccines does not prevent transmission of the virus, i.e., the vaccine is not sterilizing. However, it does reduce the amount of virus shed in the dander, hence reduces horizontal spread of the disease. Marek's disease does not spread vertically. Before the development of the vaccine for Marek's disease, Marek's disease caused substantial revenue loss in the poultry industries of the United States and the United Kingdom. The vaccine can be administered to one-day-old chicks through subcutaneous inoculation or by *in ovo* vaccination when the eggs are transferred from the incubator to the hatcher. *In ovo* vaccination is the preferred method, as it does not require handling of the chicks and can be done rapidly by automated methods. Immunity develops within two weeks.

However, because vaccination does not prevent infection with the virus, Marek's is still transmissible from vaccinated flocks to other birds, including the wild bird population. The first Marek's disease vaccine was introduced in 1970. The disease would cause mild paralysis, with the only identifiable lesions being in neural tissue. Mortality of chickens infected with Marek's disease was quite low. Decades after the first vaccine was introduced, current strains of Marek Virus cause lymphoma formation on throughout the chicken's body and mortality rates have reached 100% in unvaccinated chickens. The Marek's disease vaccine is a leaky vaccine, which means that only the symptoms of the disease are prevented. Infection of the host and the transmission of the virus are not inhibited by the vaccine. This contrasts with most other vaccines, where infection of the host is prevented. Under normal conditions, highly virulent strains of the virus are not selected. A highly virulent strain would kill the host before the virus would have an opportunity to transmit to other potential hosts and replicate. Thus, less virulent strains are selected. These strains are virulent enough to induce symptoms but not enough to kill the host, allowing further transmission. However, the leaky vaccine changes this evolutionary pressure and permits the evolution of highly virulent strains. The vaccine's inability to prevent infection and transmission allows the spread of highly virulent strains among vaccinated chickens. The fitness of the more virulent strains are increased by the vaccine.

The evolution of Marek's disease due to vaccination has had a profound effect on the poultry industry. All chickens across the globe are now vaccinated against Marek's disease (birds hatched in private flocks for laying or exhibition are rarely vaccinated). Highly virulent strains have been selected to the point that any chicken that is unvaccinated will die if infected. Other leaky vaccines are commonly used in agriculture. One vaccine in particular is the vaccine for avian influenza. Leaky vaccine use for avian influenza can select for virulent strains which could potentially be transmitted to humans.

HISTOMONIASIS

Large, pale areas in the liver of a bird infected with Histomonas meleagridis.

Histomoniasis is a commercially important disease of poultry, particularly of chickens and turkeys, due to parasitic infection of a protozoan, *Histomonas meleagridis*. The protozoan is transmitted to the bird by the nematode parasite *Heterakis gallinarum*. *H. meleagridis* resides within the eggs of *H. gallinarum*, so birds ingest the parasites along with contaminated soil or food. Earthworms can also act as a paratenic host.

Histomonas meleagridis specifically infects the cecum and liver. Symptoms of the infection include lethargy, reduced appetite, poor growth, increased thirst, sulphur-yellow diarrhoea and dry, ruffled feathers. The head may become cyanotic (bluish in colour), hence the common name of the disease, blackhead disease; thus the name 'blackhead' is in all possibility a misnomer for discoloration. The disease carries a high mortality rate, and is particularly highly fatal in poultry, and less in other birds. Currently, no prescription drug is approved to treat this disease.

Poultry (especially free-ranging) and wild birds commonly harbor a number of parasitic worms with only mild health problems from them. Turkeys are much more susceptible to getting blackhead than are chickens. Thus, chickens can be infected carriers for a long time because they are not removed or medicated by their owners, and they do not die or stop eating/defecating. *H. gallinarum* eggs can remain infective in soil for four years, a high risk of transmitting blackhead to turkeys remains if they graze areas with chicken feces in this time frame.

Symptoms

Symptoms appear within 7–12 days after infection and include depression, reduced appetite, poor growth, increased thirst, sulphur-yellow diarrhoea, listlessness, drooping wings, and unkempt feathers. Young birds have a more acute disease and die within a few days after signs appear. Older birds may be sick for some time and become emaciated before death. The symptoms are highly fatal to turkeys, but effect less damage in chickens. However, outbreaks in chickens may result in high morbidity, moderate mortality, and extensive culling, leading to overall poor flock performance. Concurrence of *Salmonella typhmurium* and *E. coli* was found to cause high mortality in broiler chickens.

Cause

A protozoan *H. meleagridis* is responsible for histomoniasis of gallinaceous birds ranging from chickens, turkeys, ducks, geese, grouse, guineafowl, partridges, pheasants, and quails. The protozoan parasite is transmitted through the eggs of a nematode, *Heterakis gallinarum*. The eggs are highly resistant to environmental conditions, and *H. meleagridis* is, in turn, highly viable inside the eggs, even for years. Birds are infected once they ingest the eggs of the nematode in soil, or sometimes through earthworms which had ingested the egg-contaminated soil. Outbreak can occur rapidly from the heavily infected bird in a flock readily through normal contact between uninfected and infected birds and their droppings in the total absence of cecal worms. For this reason, infection can spread very quickly. Once inside the digestive system of the host, the protozoan is moved to the cecum along with the eggs of *H. gallinarum*.

Transmission and Pathology

The disease causing agent, *Histomonas meliagridis*, is transmitted in the eggs of the worm *Heterakis gallinarum*. Once in the environment, the eggs are carried by earthworms. When the worms are eaten and the eggs hatch in the ceca, the pathogen is released. Bird to bird transmission can also occur from cloacal drinking.

Visible signs of this disease are cyanosis of the head (hence, "blackhead") and sulfur-yellow diarrhea. The pathogen causes lesions on the ceca and the liver. The ceca experience ulcerations, enlargement, and caseous masses start to form inside of them. The liver develops round, haemorrhagic, 1-2 centimeter oci that have caseous cores.

Diagnosis

Histomoniasis is characterized by blackhead in birds. *H. meleagridis* is released in the cecum where the eggs of the nematode undergo larval development. The parasite migrates to the mucosa and submucosa where they cause extensive and severe necrosis of the tissue. Necrosis is initiated by inflammation and gradual ulceration, causing thickening of the cecal wall. The lesions are sometimes exacerbated by other pathogens such as *Escherichia coli* and coccidia. Histomonads then gain entry into small veins of the blood stream from the cecal lesions and migrate to the liver, causing focal necrosis. Turkeys are noted to be most susceptible to the symptoms in terms of mortality, sometimes approaching 100% of a flock. Diagnosis can be easily performed by necropsy of the fresh or preserved carcass. Unusual lesions have been observed in other organs of turkey such as the bursa of Fabricius, lungs, and kidneys.

Prevention and Treatment

Currently, no therapeutic drugs are prescribed for the disease. Therefore, prevention is the sole mode of treatment. This disease can only be prevented by quarantining sick birds and preventing migration of birds around the house, causing them to spread the disease. Deworming of birds with anthelmintics can reduce exposure to the cecal nematodes that carry the protozoan. Good management of the farm, including immediate quarantine of infected birds and sanitation, is the main useful strategy for controlling the spread of the parasitic contamination. The only drug used for the control (prophylaxis) in the United States is nitarsone at 0.01875% of feed until 5 days before marketing. Natustat and nitarsone were shown to be effective therapeutic drugs. Nifurtimox, a compound with known antiprotozoal activity, was demonstrated to be significantly effective at 300–400 ppm, and well tolerated by turkeys.

POULTRY HEALTH MANAGEMENT

The best fed and housed stock with the best genetic potential will not grow and produce efficiently if they become diseased or infested with parasites. Therefore good poultry health management is an important component of poultry production. Infectious disease causing agents will spread through a flock very quickly because of the high stocking densities of commercially housed poultry.

For poultry health management to be effective a primary aim must be to prevent the onset of disease or parasites, to recognise at an early stage the presence of disease or parasites, and to treat all flocks that are diseased or infested with parasites as soon as possible and before they develop into a serious condition or spread to other flocks. To be able to do this it is necessary to know how to recognise that the birds are diseased, the action required for preventing or minimising disease and how to monitor for signs that the prevention program is working.

Principles of Health Management

The key principles of poultry health management are:

- Prevention of disease.
- Early recognition of disease.
- Early treatment of disease.

As much as is possible disease should be prevented. It is easier and less damaging to prevent disease than it is to treat it. However, it must not be assumed that all disease can be prevented. Inevitably, some will get past the defenses, in which case it becomes imperative that the condition be recognised as early as possible to allow treatment or other appropriate action to be implemented as soon as possible to bring the situation under control to limit damage to the flock.

Causes of Infectious Disease

Organisms and microorganisms that have the potential to cause harm, such as disease in animals, are called pathogens or disease vectors. There are many different types of pathogens that may be transferred from one bird to another or from one flock to another by many different means. These pathogen types include:

- Viruses.
- Bacteria.
- Fungi.
- Protozoa.
- Internal parasites.
- External parasites.

Viruses

Viruses are the smallest pathogens and can only be seen through an electron microscope. Viruses consist of an outer layer/s surrounding special protein material similar to the genetic material of the cells they invade. They can multiply and do harm only when inside the animal cell and if they invade and damage enough cells, the animal can show signs of that infection.

Antibiotics and other medications as a rule do not affect viruses and, as a consequence, there are very few medications that can treat diseases caused by viruses, although there are times when a drug may be used to control secondary infections. The best way to manage diseases caused by viruses is by quarantine and good hygiene to lower the challenge, and vaccination to maximise the birds' immunity to future challenges. Some have the ability to survive for very long periods of time in the bird dander and feather debris, litter and manure, insects and rodents.

Bacteria

Bacteria are single cell organisms with a nucleus and multiply by simple fission, which means that one divides into two, and some can do this very quickly inside the host or in a suitable environment. Some are very fragile and do not survive long outside of the host while others may survive for long periods even in a harsh environment. Many have the ability to turn into spores by forming a very tough wall that protects them from most of the materials used to kill them. These types of bacteria are much more susceptible to these compounds when not in the spore form.

Bacteria may be described as being gram positive or gram negative. This characteristic is to do with differences in their cell walls that affects their staining for viewing under the microscope. Whether they are one or the other also influences their response to certain chemicals, including disinfectants.

Different types of bacteria harm the birds in two predominant ways:

1. Those that attack and damage the birds' cells or spaces between the cells.

2. Those that produce toxins or poisons that harm the birds.

There are several antibiotics and other drugs that are effective against different bacteria. However, quarantine and good hygiene that lower the numbers to be targeted by the drugs are the important first lines of defense against these organisms.

Chlamydia

Chlamydia are a little larger than viruses. They live inside the cells they infect particularly in the cells of the respiratory system. They can be treated with antibiotics.

Mycoplasmas

These are single cell organisms slightly larger than chlamydia. They have a cell wall and nucleus. The most commonly known disease caused by this organism is Mycoplasmosis or Chronic Respiratory Disease (CRD) caused by Mycoplasma gallisepticum. Diseases caused by Mycoplasma organisms respond to some antibiotics. These organisms do not survive long outside of the host and good quarantine and hygiene procedures coupled with a suitable house de-population period will provide good control.

Fungi

Fungi are organisms larger than bacteria and are considered to be members of the plant kingdom. They multiply by forming spores that are released and enter the local environment. When conditions are satisfactory the spores start to grow to repeat the cycle.

Fungi harm the birds in two ways:

1. By being taken into the body e.g. in the respiratory system where they start to grow.

2. By producing toxins or poisons e.g. in the food. When the birds consume the contaminated food the toxin affects them. A good example of this type of damage is aflatoxin produced by certain moulds or fungi that commonly grow in peanut meal and some litter materials. Moulds or fungi are resistant to nearly all antibiotics.

Protozoa

Protozoa are single cell organisms larger than bacteria. Protozoa have a complex re-production system that, in many cases, allows them to multiply into extremely high numbers very quickly. A good example of protozoan diseases is coccidiosis of poultry.

Protozoa generally harm the birds by destroying tissue. A number of chemicals have been developed that can be used to treat birds infected by the different protozoans. Others have been developed that interfere with the protozoan life cycle and may be used as preventive treatments while the birds develop a natural immunity. These pre-ventive drugs are often referred to as coccidiastats.

Internal Parasites

Parasites are organisms that live off the host. Internal parasites in poultry are multi-celled organisms that live inside the bird usually located in specific organs. Most in-ternal parasites, and particularly those found in Australia, are visible to the naked eye.

While there are many different internal parasites found in poultry, only three are likely to cause harm. These are:

- Large round worms.

- Caecal worms.

- Tape worms.

External Parasites

These parasites live outside of the bird. Some spend all of their life on the bird while others spend only some time on the birds. Some cause harm by irritating the bird while others are bloodsuckers that, in sufficient quantity, will cause anaemia. Some of the bloodsuckers often carry organisms called spirochaetes that they inject into the bird while feeding. The spirochaetes may cause harm and tick fever is a good example that can kill many birds.

Prevention of Disease

This aspect of poultry management must receive constant, close attention. Failure to maintain a high standard will usually result in an unhealthy flock. The basis of poultry health management is:

- The isolation of the flock from disease causing organisms – quarantine.

- The destruction of as many harmful organisms as possible – hygiene.

- The use of an appropriate vaccination program – trigger the birds' immune system.

- The use of appropriate preventive medication programs – for diseases for which there are no vaccines.

- The use of a suitable monitoring program – to monitor for the presence of disease organisms and the success or failure of the hygiene program or the vaccination program.

Quarantine

The principle need is to maintain control over the means of entry by disease causing organisms. These may enter by several routes:

- Poultry: Introducing stock as day old chickens is considered to be the lowest risk method of restocking a poultry farm. Older birds are more likely to be diseased or at least carriers of disease, even if not showing signs.

- Wild birds/other animals: These often carry the causes of disease and are likely to fly or move from one poultry farm to another if the farms are close enough. The best way to prevent this is to ensure a suitable distance between farms and a minimum of 5 km is recommended. A security fence 2 metres high and with a controlled entry gate should surround the poultry farm and all sheds should be protected from entry by wild birds and all other animals by secure wire netting.

- Wind: Insects and dust carried on the wind from infected to clean farms may also carry the causal organisms of infectious disease. The best way to prevent this is to ensure a suitable distance between farms and a minimum of 5 km is recommended. This distance is influenced by the direction of the prevailing wind. Insects and dust travel further with the wind than against it, and the presence or absence of barriers in the form of hills and high vegetation that catch the dust or insects.

- People and vehicles: The most common visitors, including vehicles, are very likely to be those that have had contact with other poultry whether they be chicken delivery vehicles, feed delivery vehicles, service people and their vehicles or neighbours in the same business. Entry should only be given to essential visitors and people and vehicles should enter only through a disinfectant wash facility and visitors through a shower/change facility. Disinfectant footbaths and a change of footwear prior to entry to each shed are also recommended. In some circumstances a shower and change of clothing should be required prior to entry to all poultry house. The organisation of staff around the farm is also of importance. Wherever possible, staff should be restricted to one location. However, in some situations there is a need for staff to move from one shed to another. In these cases the principle requirement is to do so in a way that carries the least risk. This means that the normal practice is to move from youngest to oldest flocks on the farm, leaving disease flocks, no matter their age till last.

- Used equipment: No used equipment should be allowed entry to a poultry farm. If it becomes necessary to allow such entry or to move equipment from one house to another, it should be thoroughly cleaned and disinfected prior to doing so.

- Food and water: When a diseased bird eats or drinks from a trough it will leave behind contaminated food or water. While it is difficult to prevent this within one pen, if possible, the choice of feeder and drinker may minimise or slow down the transfer of disease from one bird to another. Under no circumstances should open feeders and drinkers extend from one pen to another. All drinkers and feeders should be kept clean even if they have to be cleaned daily.

- Flies and rodents: In addition to the points raised in relation to distance from other flocks to minimise the movement of insects and animals from one farm to another, all fly and rodent populations should be controlled because they can carry disease causing organisms and pass them on to the stock.

Hygiene

The practice of good hygiene kills microorganisms, including those that cause disease, and all farms carry populations of microorganisms. Therefore, good hygiene practices are an important part of poultry health management. There is an overlapping in the use of the terms quarantine and hygiene.

Good hygiene practices include:

- The thorough cleaning of poultry houses and equipment after each flock has been removed.

- The use of vehicle disinfection and wash facilities.

- The use of foot baths at the entry to each house.

- The provision of footwear at the entry to each shed.

- The use of clean litter material after washing the shed and not re-using litter. Litter in the poultry house should be managed to maintain it in a dry friable state without caking or being too wet.

- Removing all dead birds daily and disposing them in a recommended manner.

- Maintaining all houses and ancillary buildings and surrounds in a clean and tidy state.

Resisting Disease

There are a number of factors that influence whether a bird will succumb to a disease. These include:

- Genetic resistance of the birds: some genotypes are more resistant than others to infection generally while there are those that are more resistant or susceptible to specific diseases. For example, there are significant differences between at least some genotypes in their resistance to Marek's Disease.

- State of well-being of the birds in the flock: birds that are well fed and managed and kept in general good health will have a high level of well-being. Such birds are more likely to fend off an infection than those that have a low level of well-being. The immune system of unthrifty birds is usually significantly weakened.

- Level of stress in the flock: stress in a poultry flock may be caused by many situations including overcrowding, environment extremes, poor quality food and nutritional deficiencies, harassment and failure by shed staff to react in a timely manner to changing situations in the house. Stress reduces the ability of the bird to fight infection by weakening the immune system.

- The challenge or numbers of infectious organisms in the bird's environment: the greater the number or virulence (strength) of the micro-organisms the more likely they are to defeat the birds' defence and result in a disease. Quarantine and hygiene are the main ways that the number of potentially harmful micro-organisms are kept as low as possible.

- The level of immunity the birds have: this determines how well the bird can fight invasion by specific infectious organisms. Whether a bird will succumb to an infectious disease depends on the relationship between the number of infectious organisms in the environment and the level of immunity in the bird. The function of the immune system is to defend the bird against invasion by specific infectious organisms.

Many disease outbreaks only occur because there are predisposing circumstances that ensure the success of the invasion by the causal organisms. Stress in the flock is a major factor in this regard. A high level of stress reduces the bird's ability to fight the invasion by disease causing organisms. Stress, in this regard, may be environment extremes, overcrowding, nutrient deficiencies (even marginal deficiencies), infection, harassment or any other factor with the potential to stress the bird.

In some situations, a primary infection may reduce the ability of the bird to fight invasion by other organisms called secondary invaders or subsequent invaders. In many cases the bird is able to live without harm with the secondary invaders until such time their defences are lowered by the primary infection. It is in this situation that the secondary invaders cause serious harm.

A good example of this situation is the disease, collibacillosis. Collibacillosis is caused by a bacteria called Escherichia coli which is endemic in the environment and, provided normal standards of hygiene are practised and the bird is well nourished and managed, causes no real harm. However, quite often, an invasion by the organism Mycoplasma gallisepticum opens the way for the E. coli bacteria to become virulent, or for more virulent strains to gain entry resulting in the disease collibacillosis.

Disease Severity

From the point of view of flock health management, disease in poultry may be one of two levels of severity:

- Sub-clinical: A sub-clinical disease is one where the signs are not obvious. The birds do not appear to be sick but the infection causes slower growth and/or lower egg production. Sub-clinical disease may predispose to secondary invasion by other organisms. The only evidence that the birds are infected is the lower production efficiency found on an analysis of performance. In many cases this is not found until much of the financial damage has been done.

- Clinical: A clinical disease is one where the signs that the birds are sick are more obvious. They show the clinical signs typical of the disease with which they are infected. Clinical disease not only affects the performance of the flock but, in many cases, a number of the birds die or never recover to their previous performance level and remain unthrifty.

In each case, affected birds and in many cases recovered birds, are carriers that may be a source of infection for other stock with which they have contact and may transfer the causal organism either directly or indirectly to other stock not involved in this particular outbreak.

Preventative Medication

Vaccines are not available to combat all disease threats. It may be necessary to use a preventive medication to combat infection by some organisms. Veterinary advice may be necessary to determine an appropriate preventive medication program.

Monitoring Program

It is not possible to see most infectious agents. Therefore, it is appropriate to have a monitoring program. This may consist of:

- Daily checks of the flock.

- Regular on-farm and laboratory autopsies.

- Blood sampling for laboratory analysis.

- Exposing plates and taking swabs for laboratory analysis.

These techniques can be used to monitor the current disease situation including the presence of parasites, the success or failure of cleaning procedures and the success or failure of vaccinating procedures.

Recognising Healthy and Sick Birds

A very important skill for all poultry stock persons to have is the ability to differentiate between healthy and sick birds. It is normal when a flock is diseased to find healthy birds and those with varying degrees of illness. Therefore, it is necessary to be able to tell as early as possible when some of the birds in the pen are sick.

While the manager may be able to identify some diseases from available evidence, it is unlikely that they will be able to identify all. However, the sooner a disease is noticed in the flock, the sooner appropriate action can be taken. This may include initiation of a medication program, send specimens to the laboratory for examination and diagnosis, and/or to call in expert advice. Early action not only gives the manager a chance to cure the condition but it may help prevent it spreading to other stock.

A healthy bird will:

- Be active and alert.

- Be normal size/weight for the strain, age and sex.

- Have no lameness or paralysis.

- Have no injuries.

- Have no deformities.

- Have no discharges from the nostrils or eyes.

- Have no stained feathers around the vent.

- Have no swellings.

- Generally have good plumage related to the whether in a moult or in lay.

A sick bird will show some or all of the following signs:

- Isolation – sick animals usually seek a quiet place out of the way of others.

- Hunched stance – sick birds often have a hunched stance with ruffled feathers and eyes partially closed.

- Diarrhea – usually evidenced by stained feathers in the vent region.

- Paralysis – of the leg(s), wing(s) or neck.

- Sneezing and/or coughing; there may be nasal discharge.

- Blood in the faeces.

- Swellings – of the joints.

- Injuries.

- Loss of weight – this may be pronounced if the condition is a chronic one e.g. Marek's Disease.

- Unexpected changes to the food and water consumption – often the first signs of illness.

- Slower growth or a drop in egg production.

It is normal to find a continuous low level of mortality and a small number of unthrifty birds in the poultry flock. Even though this does occur, attempts should be made to reduce even this "normal" mortality. Normal levels will be in the vicinity of 4% to 40 days for meat chickens, less than 5% to point of lay for layer and breeder replacements and less than 1% per 28 days for layers and breeders.

Vaccination

Vaccination plays an important part in the health management of the poultry flock. There are numerous diseases that are prevented by vaccinating the birds against them. A vaccine helps to prevent a particular disease by triggering or boosting the bird's immune system to produce antibodies that in turn fight the invading causal organisms.

A natural invasion that actually causes the disease will have the same result as the bird will produce antibodies that fights the current invasion as well as to prevent future invasions by the same causal organisms. Unfortunately birds that become diseased usually become unthrifty, non-productive or even die. An infection caused by natural invasion will be uncontrolled and therefore has the possibility of causing severe damage, however vaccination provides a way of controlling the result with minimal harm to the birds.

Vaccines are generally fragile products, some of which are live but in a state of suspended animation. Others are dead. All have a finite life that is governed by the way they are handled and used. Handling and administration procedures also influence the potency of many vaccines and consequently the level of immunity the bird develops.

Scientists are developing a vaccine for avian influenza (Birds showing symptoms).

Types of Vaccine

Live vaccine: The active part of the vaccine is the live organism that causes the disease. As such, it is capable of inducing the disease in birds that have not had previous contact that organism. Vaccinated birds, in many cases are able to infect non-vaccinated birds if housed together.

Attenuated vaccine: With this type of vaccine the organism has been weakened by special procedures during manufacture so that it has lost its ability to cause the serious form of the disease. At worst, the birds may contract a very mild form of the disease, however, the vaccine still has the ability to trigger the immune system to produce antibodies.

Killed vaccine: With this type of vaccine the organism has been killed and is unable to cause the disease, although the ability to trigger the immune system remains. In many cases, the level of immunity produced by this form of vaccine is weaker than that produced by live and attenuated vaccines.

Vaccine Production

Vaccines are produced mainly in three forms:

1. Liquid vaccine: It is in fluid form ready to use.
2. Freeze dried vaccine: The vaccine is stored as one pack of freeze dried material and one pack of diluent, often a sterile saline solution. These have to be combined before use.
3. Dust: Where the vaccine is prepared for administration in the dry form.

Vaccines are sold in dose lots, the number of doses being the number of fowls that may be vaccinated with that amount of vaccine when using the recommended technique. In the case of many vaccines there are differences in the disease organism strains that they are effective against. It is important that the correct vaccine strain be used and this can only be determined by veterinary advice.

Handling Vaccines on the Farm

Vaccines are fragile in many respects and require very careful handling to ensure they retain their potency. Poor handling procedures will, in most cases, result in a rapid decline of potency.

The important handling requirements on the farm are:

On receipt of the vaccine on the farm, check and record:

1. That the vaccine has been transported in the recommended manner which is usually in the chilled or frozen state. Prolonged exposure to atmospheric temperature will result in rapid loss of potency.

2. Type of vaccine: Is it the vaccine ordered.

3. The number of doses: Has the correct amount been delivered.

4. The expiry date of the vaccine: Vaccines have a date by when there is a significant risk that they will no longer retain their potency and will not produce the immunity required. The expiry date is based on the vaccine being handled and stored in the recommended manner.

    •   As soon as possible place the vaccine into recommended storage conditions. Read the instructions to find out what these are. However, freeze dried material should be kept at a temperature below freezing and its diluent at a temperature just above freezing. Liquid vaccines are generally kept at temperatures just above freezing.

    •   Remove the vaccines from storage immediately prior to their being used. Only remove and re-constitute enough for immediate needs and repeat this through the day if more is required. Do not mix what is required for an entire day at the start of the day and leave it stand until required, as the vaccine will rapidly lose it efficacy.

    •   Protect the vaccines after mixing by holding them in an ice bath. Place ice in a small esky or similar container and place the container of mixed vaccine in the ice. Some vaccines have a very short life once mixed. For example, Marek's Disease has a life of about 1.5 hours after mixing if held in an ice bath. It is much shorter if held in higher temperatures.

    •   Use the recommended administration techniques and do not vary these without veterinary advice.

    •   Always clean and sterilise the vaccinating equipment thoroughly after use.

    •   Always destroy unused mixed vaccines after the task has been completed. Some vaccines have the potential to cause harm if not destroyed properly.

    •   Do not vaccinate birds that are showing signs of disease or stress.

Vaccination Procedures

There are a number of ways that vaccines may be administered to poultry and it is very important that the correct method be used for each vaccine. To use the wrong method will often result in failure of the vaccine to produce the desired immunity. Some of the methods require the operator to handle every bird and, consequently are time consuming and stressful to the birds and operator. Other methods involve administration by methods much less stressful and time consuming. These methods include administration via the drinking water or as an aerosol spray. The different ways that the vaccines may be administered to poultry are below.

In-ovo Vaccination

Using the method of in-ovo vaccination, the vaccine is administered into the embryo before hatch.

In general, vaccines can be applied to five different areas of the egg: the air cell, the allantoic sac, the amniotic fluid, the body of embryo and the yolk sac. Vaccine uptake and therefore the immune response of the chicken depend largely on the area of injection. While injection in the air cell has been shown to be minor/not effective, injection in the body of the embryo or the allantoic sac is effective. Therefore, the optimum period to inject the embryo is in the late stage of development, i.e. the time between the ascendance of the stalk of the yolk sac into the abdomen (about the time when the chicken tucks its head under its wings) and external pipping.

During that late stage of development, the embryo is mature enough to cope with the viral stimulus and the trauma induced by the penetrating needle is unlikely to cause severe tissue damage. Signs of too early vaccination include reduced hatchability, late death and increased number of culled birds. However, if vaccination is done too late in embryonation, the risk of egg shell breakage is significantly higher. Therefore, in ovo vaccination is commonly performed between days 18-19 of incubation.

The system of a larger outer needle (penetrating the egg shell) that contains an inner needle (penetrating the embryo) enables for strong but careful penetration of the egg and minimizes trauma to the embryo. In addition, the use of two needles reduces the likelihood of transferring contaminants on the outer egg shell into the sterile embryo. The needle for punching the egg shell should not penetrate the embryonic cavity (the inner shell membrane, the chorio-allantoic membrane or air cell membrane). While the penetration of the outer egg shell increased the relative pore volume about 30%, the risk for increased gas exchange of the embryo occurs.

Hygiene management including reduced air circulation, well maintained air filters, adjustment to weather conditions and well maintained hatchery insulation has to be taken into account when performing in ovo inoculation. Only strict management of these environmental factors can reduce the likelihood of infections of the egg, especially with aspergillosis or other air-borne pathogens. Continuous training of reliable staff is of highest priority to prevent reduced hatchability and to maintain high hygienic standards. A sterile environment and the usage of chlorine based sanitizers are crucial. The storage and preparation of the vaccine in a separate biosecure area as well as strict precautions in using sterile devices such as containers and water should be implemented. While the cost of machine acquisition is high, the investment can pay back by its advantages.

The advantage of commencing immunity development before hatch can prevent young chicks from early infection after hatch. Since high tech machines are used for in ovo injection, the volume and concentration of the vaccine to be administered are highly standardised, reducing human error and labour cost when compared to vaccination of chickens later in life. Furthermore, vaccination of every single chicken can be ensured resulting in better uniformity of the flock. Coming with this is improved animal welfare due to less handling of birds later in life.

Currently Marek's disease, Newcastle disease, infectious laryngotracheitis and infectious bursal disease vaccines are routinely administered using in ovo vaccination in various countries. In ovo vaccination does not interfere with maternal antibodies that may still present in the embryo. In fact, it increases the level of immunity and as a consequence one injection is sufficient to offer life-long protection against the target disease.

Intramuscular Injection

This method involves the use of a hypodermic needle or similar equipment to introduce the vaccine into the muscle (usually the breast muscle) of the bird. The task is sped up greatly by the use of an automatic syringe which makes the technique relatively easy and doesn't harm the bird. Care must be taken to ensure that the correct dose is administered to each chicken and the equipment should be checked regularly to ensure this.

Care must be taken to ensure that the needle does not pass through into a key organ and that other unwanted organisms are not administered to the bird at the same time by contaminated vaccine or equipment. Contamination can be prevented by good hygiene and vaccine handling procedures.

Subcutaneous Injection

This method involves the use of similar equipment to that used for the intramuscular technique. The main difference between the two techniques is that, in this case, the vaccine is injected under the skin, usually at the back of the neck, and not into the muscle. Care must be taken to ensure that the vaccine is injected into the bird and not just into the feathers or fluff in the case of very young chickens. The dose being administered should be checked for accuracy frequently. Maintain good hygiene practices to limit introducing contaminating organisms during the procedure.

Ocular

This method involves the vaccine being put into one of the bird's eyes. From here the vaccine makes its way into the respiratory tract via the lacrimal duct. The vaccine is delivered through an eyedropper and care must be taken to ensure that the dropper delivers the recommended dose. If it is too little, the level of immunity may be inadequate, while if too much, the vaccine may not treat the total flock but will run out beforehand.

Nasal

This method involves introducing the vaccine into the birds' nostrils either as a dust or as a drop. Always ensure that the applicator delivers the correct dose for the vaccine being used.

Oral

With this method the vaccine dose is given in the mouth. From here it may make its way to the respiratory system or it may continue in the digestive tract before entering the body.

Drinking Water

With this method the vaccine is added to the drinking water and, as a consequence, is less time consuming and is significantly less stressful on the birds and operator. Take care to ensure the vaccine is administered correctly as there is much scope for error. The recommended technique observes the following:

- All equipment used for vaccination is carefully cleaned and free of detergents and disinfectants.

- Only cold, clean water of drinking quality should be used.

- Open the stopper of vaccine bottle under water.

- The water present in the drinking trough should be consumed before vaccination.

- By ensuring that all birds drink during the vaccination phase, all should receive an adequate dose of the vaccine.

Cloacal

This method involves the introduction of the vaccine to the mucus membranes of the cloaca with an abrasive applicator. The applicator is firstly inserted into the vaccine and then into the bird's cloaca and turned or twisted vigorously to cause an abrasion in the organ. The vaccine enters the body through the abrasion. The technique is time consuming and stressful to the birds and care must be taken to ensure no contamination is introduced with the vaccine particularly from bird to bird. As a rule, the technique is not used on commercial farms.

Feather Follicle

With this method the vaccine is introduced into the feather follicles (the holes in the skin from where the feathers grow). The technique involves the removal of a group of adjacent feathers or fluff in young chickens, and the brushing of the vaccine into the empty follicles with a short, stiff bristled brush. Good hygiene is necessary to prevent the introduction of contaminant organisms with the vaccine.

Wing Stab

With this method the vaccine is introduced into the wing by a special needle(s). These needles have a groove along their length from just behind the point. When dipped into the vaccine some of the vaccine remains on the needle to fill the groove. The needle(s) are then pushed through the web just behind the leading edge of the wing and just out from its attachment to the body of the bird. Care must be taken to select a site free of muscle and bone to prevent undue injury to the bird. Ensure that the needles penetrate the layers of skin at the ideal site. A common problem is for the vaccine to be brushed from the needles by fluff or feathers before it is brushed into the follicles.

Spray

With this method the vaccine is sprayed onto the chickens (or into the air above the chickens) using a suitable atomiser spray. The vaccine then falls onto the chickens and enters the body of other chickens as they pick at the shiny droplets of vaccine. A small quantity may be inhaled as well.

Monitoring

In the case of some vaccines, an important part of the procedure is to ascertain whether the vaccine has worked, or "taken". A good example of this is fowl pox vaccine, which is administered by wing stab. Within 7 to 10 days after vaccination a "take" should appear at the vaccination site. This is in the form of a small pimple one half to one centimetre in diameter. If the take is larger and has a cheesy core, it indicates that contaminants have been introduced either with the vaccine or with dirty vaccinating equipment. A check for takes would involve inspecting approximately 100 birds for every 10,000 vaccinated.

Another example of whether the birds are reacting satisfactorily to the vaccination is the systemic reaction found in chickens vaccinated against infectious bronchitis disease. In many cases the birds react approximately 5 to 7 days after vaccination by showing signs if ill health such as slight cough, a higher temperature and lethargy. In cases where there are no obvious signs of success, blood samples may be taken and sent to the laboratory for examination. The usual test is for the presence of an adequate number of the appropriate antibodies (called the titre) in the blood. If the vaccination has been unsuccessful, it may be necessary to re-vaccinate to obtain the desired protection.

Failure to find evidence of success could be because of:

- Faulty technique resulting in the vaccine not being introduced into the vaccination site.

- Faulty vaccine – too old or not stored or mixed correctly. It would be unusual but not impossible for the vaccine to be faulty from manufacture.

- The birds are already immune i.e. the immune system has already been triggered as a result of parental (passive) immunity, previous vaccination or other exposure to the causal organism.

PREVENTION AND CONTROL OF POULTRY DISEASES

Prevention and control of poultry diseases is one of the most important factor for the profitability of poultry farming business. Following basic factors should be kept in mind for preventing and controlling of poultry diseases.

Some diseases can be controlled by both vaccination and keeping them out of farms. These include coryza; chronic respiratory disease, caused by Mycoplasma gallisepticum; infectious laryngotracheitis; lice and mite infestations; chlamydiosis; blackhead; and internal parasites.

How to Keep Diseases Out: Disease can enter your farm via carrier birds, people, wild birds, day-old chickens, equipment, wind, pets and insects.

Birds: Apparently healthy birds carrying a disease organism can be a source of infection of other birds. If disease-carrying started pullets are introduced onto an uninfected farm they can spread disease. Backyard, show and aviary birds can also carry disease.

Prevention: Do not keep backyard, aviary, show birds or other birds such as emus on commercial poultry farms. Make sure that you, your employees and visitors to your sheds have not had any contact with these birds. Do not keep domestic ducks on poultry farms, other than duck farms. Purchase your started pullets from reputable suppliers where the disease status is known.

People: People are probably the second most common carrier of poultry diseases. Disease can be carried on footwear, hands, clothing and possibly in the nostrils. Visitors from overseas could spread exotic diseases. Poultry producers, family or staff members can bring disease back onto farms.

Prevention: Do not allow people onto your farm unless they have some essential task to perform. To safeguard the health of your flocks, make sure that contract work crews, service people and veterinarians who enter sheds take stringent precautions such as washing their hands and changing their overalls and shoes before entering your sheds. This applies particularly to visitors who have been on other poultry farms that day. The poultry farm should be surrounded by a security fence and have a single gateway fitted with a 'Restricted Access' sign. Do not allow people who are picking up eggs, or sales and feed delivery personnel, to enter sheds. If your birds are kept on the floor there is a risk of spreading disease if you wear the same pair of boots into different sheds. Keep a separate set of boots for wearing in each shed and store them in a receptacle outside the door.

Wild Birds: A surveys in Australia indicate that a very small percentage of waterfowl are infected with avian influenza (AI) viruses. The H5N1 AI virus which is currently causing problems around the world has not been found in Australia. Water carrying these viruses is thought to be responsible for some avian influenza outbreaks. Pigeons contaminating feed in the United Kingdom in 1984 caused 23 cases of Newcastle disease. Carrion-eating birds such as crows can spread disease in free-range enterprises. Wild birds can also spread external parasites.

Prevention: Great effort is warranted to make sure that wild birds, especially waterfowl, cannot enter sheds. Bird-proof your sheds, and shut the doors when the sheds are not being attended. Install plastic hanging strips to deter birds while the sheds are being used. Discourage waterfowl from coming close to sheds by cleaning up feed spillages promptly and draining wet areas near sheds. Make sure that water for drinking and fogging is not contaminated by free-flying birds. Chlorination or ultraviolet treatment is recommended for all dam or river water and this should be combined with suitable water filtration. Make sure that all water tanks are covered adequately and that feed is not contaminated by wild birds, animals or vermin.

Day-Old Chickens: Egg-borne disease can be transmitted from the infected hen to the day-old chicken via the fertile egg. Two examples are:

- Chronic respiratory disease, caused by Mycoplasma gallisepticum.

- Infectious synovitis, caused by Mycoplasma synoviae.

Prevention: Day-old chickens can be bought free of M. gallisepticum.

Equipment: Diseases can be introduced on equipment which is shared between farms. Poultry crates and fibrous egg flats can be transmitters of disease organisms. Avian influenza, EDS 76, Newcastle disease and northern fowl mite, among others, can be transmitted from farm to farm on egg flats.

Prevention: Do not share equipment between farms and do not use second-hand egg fillers.

Wind Spread: Some diseases, particularly respiratory diseases, can be blown in the air from one farm to another. This commonly occurs at night or on cloudy days when the sun's ultraviolet radiation is not present to kill the infective agent. There is evidence that chronic respiratory disease, Mycoplasma gallisepticum, avian influenza and Newcastle disease may be spread this way.

Prevention: Keep poultry farms as far apart as possible if you are setting up a new farm. Sheds should also be built as far from the road as possible. Trees growing between farms and between the farm and the road will break up wind movements.

Pets: Dogs and cats can carry infectious material on their feet and coats and can put your birds at risk if they visit neighbouring farms or dead-bird-disposal areas.

Prevention: Secure your poultry sheds against the entry of dogs and cats. Keep the doors closed when your sheds are not being serviced.

Insects: Mosquitoes can transmit fowl pox, and flies can spread some species of tapeworm, Newcastle disease and salmonella.

Prevention: Vaccinate all birds against fowl pox if mosquitoes are a problem, and reduce the number of flies.

How to stop diseases spreading: The following procedures won't stop diseases getting into farms, but they will stop them spreading and reduce their severity:

- Ensure all birds are correctly vaccinated and medicated. Follow a suitable vaccination regimen for the diseases that occur in your area. Use and care for your vaccines as directed on the label. Preventative medications (for example coccidiostats) may be necessary for some conditions. Vaccination against Newcastle disease is compulsory in NSW and most other states.

- Have one age of bird per farm. Having one age of bird per farm allows any acquired diseases to be eradicated. Make sure that incoming started pullets and day-old chickens are free of disease and that strict quarantine procedures are in place on the farm. After the batch of birds is sold, clean the sheds and equipment thoroughly and allow 2 weeks (the depopulation period) before bringing in the next batch.

- Use all-in all-out sheds. If it is not practical to have only one age of bird on the farm, reduce the number of age groups to a minimum. If you have fewer age groups than sheds (for example if you have four sheds and three age groups), try to have the same age group in the sheds that are closest together. Egg packers and other workers should preferably be allocated specific sheds to work in. If this is not possible and they have to go into all the sheds, the general direction of movement should be from the youngest birds to the oldest.

- Dispose of dead birds properly. Dead birds should be quickly burnt, deeply buried or effectively composted and should never be fed to cats or dogs. Dead birds left lying around the farm can spread disease to other sheds and neighbouring farms via carrion-eating birds,dogs, cats and rats. Recapture escaped birds. Recapture escaped birds quickly. If a bird has been free for an undetermined length of time and has got out of the shed, it should not be returned to the main flock. The bird-proofing recommended to stop wild birds getting into your sheds will also stop escaped birds from getting out.

- Inspect your farm daily. Finally, inspect your sheds daily so that any problems can be identified early and rectified quickly. This will minimise the degree of poultry diseases challenge.

ISSUES AT THE LEVEL OF PRODUCTION AND PROCESSING UNITS

Animal Production Units

Local disturbances (e.g. odour, flies and rodents) and landscape degradation are typical local negative amenities in the surroundings of poultry farms. Pollution of soil and water with nutrients, pathogens and heavy metals is generally caused by poor manure-management and occurs where manure is stored. Water and soil pollution related to poultry litter is, however, generally not an issue at the production site, as poultry manure is only directly discharged into the environment in exceptional conditions. Indeed, the high nutrient content and low water content of poultry litter make it a valuable input to agriculture. Manure is either recycled on cropland belonging to the animal farm or marketed. In the usual set-up, an intermediary or a processor collects manure from poultry farms. Manure is either resold rough or processed into compost or pellets. Manure products are used as fertilizer, or as animal feed especially for fish and cattle. In south Viet Nam, the authors observed that end users may be located as far as 300 km from the animal farm where manure is produced. An intermediary will sell manure to the group of users with highest willingness to pay, which can change throughout the year, and from year to year, according to the cropping calendar and the economic conditions. Manure price at the animal-farm gate varies with its pureness (presence of litter) and water content and with the season (demand). On average, 20 kg bags of fresh chicken manurewithout litter are sold for VND4 000 to 6 000 while 20 kg bags of manure with litter are sold for VND1 500 to 2 000.

Local Disturbances

Poultry facilities are a source of odour and attract flies, rodents and other pests that create local nuisances and carry disease. Odour emissions from poultry farms adversely affect the life of people living in the vicinity. Odour associated with poultry operations comes from fresh and decomposing waste products such as manure, carcasses, feathers and bedding/litter. On-farm odour is mainly emitted from poultry buildings, and manure and storage facilities. Odour from animal feeding operations is not caused by a single compound, but is rather the result of a large number of contributing compounds including ammonia (NH_3), volatile organic compounds (VOCs), and hydrogen sulphide (H_2S). Of the several manure-based compounds which produce odour, the most commonly reported is ammonia. Ammonia gas has a sharp and pungent odour and can act as an irritant when present in elevated concentrations. Odour is a local issue, which is hardly quantifiable; the impact greatly depends on the subjective perception of populations neighbouring the farm. It is, therefore, difficult to evaluate the maximum distance over which odorous gas travels; however, odour problems are generally concentrated within 500 metres of the farm. Although generally not causing any public-health concern, odours can represent a strong local problem that is frequently reported by farms' neighbours as the most disturbing environmental impact. The emission of odours mostly depends on the frequency of animal-house cleaning, on the temperature and humidity of the manure, on the type of manure storage, and on air movements. For these reasons it is generally higher in waterfowl farms than in chicken farms. Flies are an additional concern for residents living near poultry facilities. Research conducted by the Ohio Department of Health indicated that residences that were located in close proximity to poultry facilities (within half a mile) had 83 times the average number of flies. In addition to the nuisance they cause, flies and mosquitoes can transmit diseases, such as cholera, dysentery, typhoid, malaria, filaria and dengue fever. Although less often reported than flies and mosquitoes, rats and similar pests are also a local nuisance associated with poultry production. As with flies and mosquitoes, they can be a vector for disease transmission. Their presence is mainly related to animal-feed management and especially to storage and losses from feeding systems. Pesticides used to control pests (e.g. parasites and disease vectors) and predators have been reported to cause pollution when they enter groundwater and surface water. Active molecules or their degradation products enter ecosystems in solution, in emulsion or bound to soil particles, and may, in some instances, impair the uses of surface waters and groundwater.

Land use and Landscape

The trend to larger production units, and their regional concentration, certainly has the potential to adversely affect surrounding land use and the appearance of the landscape. Massive industrial poultry installations can create an adverse aesthetic impact. Impact on land use in highly concentrated areas is manifested through conflict with development needs and in some areas with rural tourism.

Poultry Carcass Disposal

Improper disposal of poultry carcasses can contribute to water-quality problems especially in areas prone to flooding or where there is a shallow water table. Methods for the disposal of poultry carcasses include burial, incineration, composting and rendering. In the case of recent highly pathogenic avian influenza (HPAI) outbreaks, the disposal of large numbers of infected birds has presented new and complex problems associated with environmental contamination. Large volumes of carcasses can generate excessive amounts of leachate and other pollutants, increasing the potential for environmental contamination. Buried birds undergo a decomposition process. During this process, nutrients, pathogens and other components of the carcass are released into the environment. As these substances enter the surrounding soil, they may be broken down, transformed, lost to the air, or otherwise immobilized so that they pose no environmental threat. However, there is a possibility that some constituents may eventually contaminate soil, groundwater and surface water. Another related problem is the removal of manure from houses that contain infected birds. Ritter et al. examined the impact of dead-bird disposal on groundwater quality. They monitored groundwater quality around six disposal pits in Delaware. Producers in Delaware were using open-bottomed pits for their day-to-day mortality disposal. These pits are not strictly the same as burial pits, though there are some similarities. Most of these pits were located in sandy soils with high seasonal water tables. The potential for pollution of groundwater is high with this method of disposal. After selecting the sites, two to three monitoring wells were placed around each pit to a depth of 4.5 metres. Ammonia concentrations were high in two of the wells. Three of the disposal pits caused an increase in ammonia concentrations in the groundwater. Total dissolved solids concentrations were high in all monitoring wells for most dates. Bacterial contamination of groundwater by the disposal pits was low.

Slaughterhouse

The most significant environmental issue resulting from slaughterhouse operations is the discharge of wastewater into the environment. Like many other food-processing activities, the necessity for hygiene and quality control in meat processing results in high water usage and consequently high levels of wastewater generation.Poultry processing activities require large amounts of high-quality water for process cleaning and cooling. Typical water usage in poultry slaughterhouses ranges between 6 and 30 cubic metres per tonne of product. Large quantities of water are consumed in poultry slaughterhouses for evisceration, cleaning and washing operations.

Process wastewater generated during these activities typically has high biochemical and chemical oxygen demand (BOD and COD3) due to the presence of organic materials such as blood, fat, flesh, and excreta. In addition, process wastewater may contain high levels of nitrogen, phosphorus, and residues of chemicals such as chlorine used for washing and disinfection, as well as various pathogens including Salmonella and Campylobacter. Poultry by-products and waste may contain up to 100 different species of micro-organisms, including pathogens, in contaminated feathers, feet and intestinal contents. Typical values for wastewater produced from poultry processing are 6.8 kg BOD per ton live weight killed (LWK) and 3.5 kg suspended solids per ton of LWK.

Poultry slaughterhouses release large amounts of waste into the environment, polluting land and surface waters as well as posing a serious human-health risk. The discharge of biodegradable organic compounds may cause a strong reduction of the amount of dissolved oxygen in surface waters, which in turn may lead to reduced levels of activity or even death of aquatic life. Macronutrients (nitrogen, phosphorus) may cause eutrophication of the affected water bodies. Excessive algal growth and subsequent dying off and mineralization of these algae may lead to the death of aquatic life because of oxygen depletion.

Slaughterhouses are usually located in urban or peri-urban locations, where transport costs to markets are minimized and where there is abundant labour supply. This situation increases the risk of environmental impacts: first, because slaughterhouses often lack the land required to set up waste-management facilities; second, because the pollutants that are emitted add to those emitted by other human activities; and third, because neighbouring communities are directly affected by surface-water and groundwater contamination.

Watershed-level Pollution Associated with Waste Management

Intensification of production and the geographical concentration of production units often results in environmental concerns. The decoupling of crop and livestock production through the migration of livestock production away from crop activities into areas with little or no agricultural land leads to high levels of environmental impact – mainly related to manure mismanagement and nutrient overloads.

Poultry Manure

Poultry manure contains considerable amounts of nutrients such as nitrogen, phosphorus, and other excreted substances such as hormones, antibiotics, pathogens and heavy metals which are introduced through feed. Leaching and runoff of these substances has the potential to result in contamination of surface water and groundwater resources.

Nutrients

Animals reared in intensive production systems consume a considerable amount of protein and other nitrogen-containing substances in their diets. The conversion of dietary nitrogen to animal products is relatively inefficient; 50 to 80 percent of the nitrogen is excreted. Nitrogen is excreted in both organic and inorganic compounds. Nitrogen emissions from manure take four main forms: ammonia (NH_3), dinitrogen (N_2), nitrous oxide (N_2O) and nitrate (NO_3^-).

Phosphorus is an essential element for animal growth. Unlike nitrogen, phosphorus is relatively stable once attached to soil particles and does not leach through the soil into groundwater. It does not pose any environmental risks except as a nutrient; it limits biological activity in water resources and builds up in soil when applied in excess. Phosphorus emissions from manure occur in one main form: phosphate (P_2O_5).

Heavy Metals

Manure contains appreciable quantities of potentially toxic metals such as arsenic, copper and zinc. In excess, these elements can become toxic to plants, can adversely affect organisms that feed on these plants, and can enter water systems through surface run-off and leaching. Trace elements are introduced into poultry diets either involuntarily through contaminated feedstuffs or voluntarily, as feed additives used to supply animals' requirements or – in much greater proportions – as veterinary medicines or growth promoters.

Drug residues antimicrobial agents are administered to poultry for therapeutic reasons or to prevent illness (prophylaxis). At much lower doses (subtherapeutic doses) antimicrobial agents are used as feed additives to increase the rate of growth and to improve feed efficiency. Irrespective of dosage, an estimated 75 percent of antimicrobial agents administered to confined poultry may be excreted back into the environment. Recent evidence suggests that the interaction between bacterial organisms and antimicrobials in the environment may contribute to the development of antimicrobial-resistant bacterial strains. In a study that evaluated the presence of antimicrobial compounds in surface water and groundwater resources proximal to intensive poultry operations in Ohio, found antimicrobial residues to be prevalent – present in 12 water samples (67 percent) collected proximal to poultry farms.

In the United States of America, overall use of antimicrobials for non-therapeutic purposes in animals rose by about 50 percent between 1985 and 2001. This was primarily driven by increased use in the poultry industry, where non-therapeutic antibiotic use increased from 2 million to 10.5 million pounds (907 185 kg to 4 762 720 kg) between the 1980s and 2001 – which amounted to a dramatic 307 percent increase on a per-bird basis.

Pathogens

Manure also contains pathogens which may potentially affect soil and water resources, particularly if poorly managed. Parasites such as Cryptosporidium and Giardia spp. can easily spread from manure to water supplies and can remain viable in the environment for long periods of time.

Regional Concentration of Production

The trend toward clustering of poultry production in certain preferred locations is ongoing in developed as well as developing economies. An analysis of hen populations at municipio level in Brazil, for example, shows an increasing concentration during the period 1992 to 2001. In 1992, 5 percent of the country's total area hosted 78 percent of the chicken population, while in 2001 the same area was home to 85 percent of the population.

Clustering is a process of geographic concentration of production units. This gives rise to groups of interconnected producers, feed mills, slaughterhouses and processing units. Clustering is driven by economies of agglomeration – the benefits that individual units obtain when they locate close to one another. Basically, the more related units clustered together, the lower the unit cost of production and the larger the market that individual units can sell into. In the livestock sector, lower production costs are achieved through competition among suppliers of inputs (e.g. feed mills, veterinary and other services), and specialization and division of labour among producers (e.g. breeding operations, fattening operations and contract farming). If a well-developed transport infrastructure supports this set-up, supply to urban and export markets is often very competitive.

Intensive production, therefore, concentrates in areas favoured by cheap inputs (particularly feed) and services, and by good market outlets for livestock products. Such conditions are found in the vicinity of cities, feed processors and large slaughterhouses, as well as harbours trading feed and animal products. The geographical location of intensive poultry activity is, thus, less and less linked to agricultural and land-use parameters. In other words, poultry production is shifting from agricultural use of the land, based on biophysical criteria (e.g. soil quality, climate, length of growing period) towards industrial use of the land.

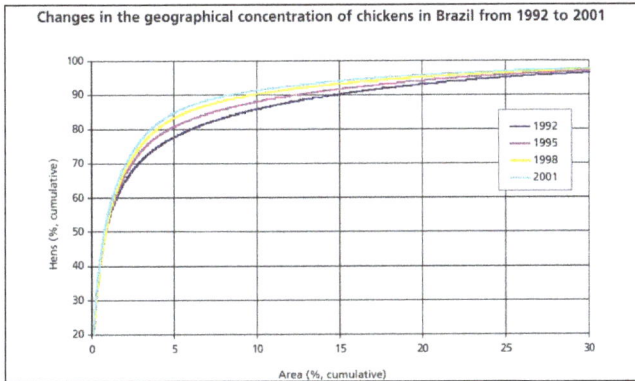

Changes in the geographical concentration of chickens in Brazil from 1992 to 2001.

Environmental Issues

Some of the major potential impacts of intensive livestock production on land and water resources:

- Eutrophication of surface waters, caused by the input of organic substances and nutrients either through wastewater from production, runoff or leakages from storage and handling facilities – affecting aquatic ecosystems and drinking water quality;

- Leaching of nitrate, and possible pathogen transfers to groundwater – affecting the quality of drinking water;

- Accumulation of nutrients and other elements in soil due to continuous application of excess quantities of manure; and

- Impacts of pollution on nutrient-sensitive ecosystems resulting in biodiversity losses. In most cases, structural changes in the production system have a rather negative impact on manure management practices. In particular, growth in the scale of production and geographical concentration in the vicinity of urban areas, cause dramatic land–livestock imbalances, hampering manure recycling options. Indeed, in such conditions, transport costs associated with carrying manure back to the field are prohibitive.

Contribution to Regional-level Nutrient Overloads

As mentioned above, poultry manure is generally recycled. Despite this apparently safe handling, it often contributes to nutrient-based pollution at regional level. First, areas where poultry production concentrates are also often characterized by high populations of other livestock species, pigs in particular. Poultry manure, thus, contributes to the structural nutrient overload in these areas. Secondly, the manure may be applied to crops or fish ponds in excess or in addition to chemical fertilizers or fish feed, resulting in an over-supply of nutrients. Such saturated systems will release excessive nutrients into the environment. Excessive levels of nitrogen in the environment lead to a cascade of effects, including:

- Decreased species diversity and acidification of non-agricultural soils, due to nitrogen deposition related to ammonia and nitrous oxide emission;

- Eutrophication of surface waters, including excess algal growth and a decrease in natural diversity due to runoff of nitrogen from agricultural soils;

- Pollution of groundwater due to nitrate leaching from agricultural soils and nonagricultural soils; and

- Greenhouse gas emissions in the form of nitrous oxide.

Nitrogen pollution has been identified as posing a risk to the quality of soil and water. These risks relate to high levels of nitrates, which can be leached to the groundwater table or to surface water causing eutrophication. In its nitrate form, nitrogen is very mobile in soil solution and can easily be leached below the rooting zone and into groundwater.

The rapid growth of intensive poultry production in many parts of the world has created regional and local phosphorus imbalances. The application of manure has resulted in more phosphorus being applied than crops require, and increased potential for phosphorus losses in surface runoff. This situation is exacerbated by manure management being nitrogen based. When manure is applied to meet the nitrogen needs of most crops, a substantial build-up of phosphorus occurs in the soil. Environmental problems associated with phosphorus losses from soils can have significant off-farm impacts on water quality. In some cases, these impacts are manifested many miles from the site where the phosphorus losses in soil erosion and runoff originally occurred. Too much phosphorus input into a body of water leads to plant overgrowth, shifts in plant varieties, discolouration, shifts in pH, and depletion of oxygen as a result of plant decomposition. A drop in the level of dissolved oxygen in surface water has deleterious effects on fish populations. Thus, increased outputs of phosphorus to fresh water can accelerate eutrophication, which impairs water use and can lead to fish kills and toxic algal blooms. In general, 80 percent of the phosphorus contained in animal feed is subsequently excreted.

Food- and water-borne diseases are another major issue associated with manure management. Pathogens are mostly transmitted through untreated animal waste. Recycling manure is a cost-effective way to reduce discharge into the environment and contamination of water systems. However, recycling must be controlled carefully in order to avoid transferring pathogens to the human food chain. Nonetheless, manure is usually not treated, even if limited composting may take place when manure is stored over several weeks (on farm or in a middleman's barn) and crop residues are added.

Soil Contamination with Heavy Metals

With increasing use of metals not only as growth promoters, but also as feed additives to combat diseases in intensive poultry production, manure application has emerged as an important source of environmental contamination with some of these metals. Metals such as arsenic, cobalt, copper, iron, manganese, selenium and zinc are added to feeds as a means to prevent disease, improve weight gain and feed conversion, and increase egg production. Typically, animals can absorb only 5–15 percent of the metals they ingest. The majority is therefore excreted in manure. Part is absorbed by the soil, but heavy metals can also end up in water bodies where they become more concentrated.

The environmental risk associated with heavy metals is largely dependent on the soil's ability to adsorb and to desorb these elements, and the potential for leaching or soil-loss to water by erosion. The spreading of animal manure contaminated with heavy metals can lead to an accumulation of these elements in agricultural soils and water bodies. Unlike excess nitrogen and phosphorus applied to land, heavy metals such as zinc and copper remain bound to soil and do not migrate to water supplies except during soil erosion. The concentrations of copper and zinc needed by animals are moderately low – 8 parts per million (ppm) for copper and 40 ppm for zinc. Yet, throughout the United States of America, most broiler diets contain levels of 125 to 250 ppm of copper in order to improve feed efficiency. The U.S. Geological Survey has reported that intensive poultry production units in the Delaware–Maryland–Virginia (Delmarva) Peninsula, on the eastern shore of the United States of America are introducing between 20 and 50 tonnes of arsenic into the environment annually.

Ecosystem Contamination

With drug residues and hormones The excretion of hormones from poultry has been cited as a possible cause of endocrine disruption in wildlife. Endocrine disruptors are a class of compounds (either synthesized or naturally occurring), which are suspected to have adverse effects in animals. They affect organisms primarily by binding to hormone receptors and disrupting the endocrine system. Endocrine disrupting chemicals (EDCs) include pesticides, herbicides and other chemicals that interact with endocrine systems.

In poultry production, EDCs can both enter and leave the production cycle. Sources of EDCs during the production phase include contaminants in litter and from grains used as feed. Poultry can also produce EDCs in the form of steroid hormones that are excreted in manure. The steroids of greatest concern are estrone and 17-ß-estradiol.

Research has shown that poultry litter contains estrogen (17-ß-estradiol), estrone and testosterone in measurable concentrations, and that these EDCs persist in the litter. Degradation of steroids in poultry litter during storage is minimal. However, once steroids have reached waterways their degradation is rapid. Research into the endocrine disruption impact of naturally occurring steroids on fish suggests that on runoff from fields where poultry manure has been applied steroid levels are high enough to cause endocrine disruption resulting in reproductive disorders in a variety of wildlife. Endocrine disruption resulting from intensive poultry production has been well documented in the Delmarva Peninsula in the United States of America.

Ecosystem Contamination through Ammonia Deposition

Atmospheric ammonia (NH_3) is increasingly being recognized as a major air pollutant because of its role in regional-scale tropospheric chemistry and its effects when deposited into ecosystems. Ammonia is a soluble and reactive gas. This means that it dissolves, for example in water, and that it will react with other chemicals to form ammonia-containing compounds. The concentrations of ammonia in the air are greatest in areas where there is intensive livestock farming. Agricultural land receiving large inputs of nitrogen from manures normally acts as a source of ammonia, but it may also act as a "sink" and absorb ammonia from the atmosphere. There is little deposition of ammonia gas to intensively managed farmland, which is largely a net source of ammonia. Ammonia in the atmosphere can be absorbed by land, water and vegetation (known as dry deposition). It can also be removed from the atmosphere by rain or snow (wet deposition). Impacts of ammonia deposition include; soil and water acidification, eutrophication caused by nitrogen enrichment with consequent species loss, vegetation damage, and increases in emissions of the greenhouse gases such as nitrous oxide.

Impacts on the Global Environment

Environmental impacts of poultry production are not always confined to specific areas; they also include impacts of a global dimension. Two issues are of relevance: the production of concentrate feed and greenhouse gas production related to energy use in animal production processes and in the transport of processed products. This topic analyses these two issues in the context of poultry production and the sector's impacts on the environment.

Feed Production

The extraordinary performance of the poultry sector over the past three decades has partially been achieved through soaring use of concentrate feed, particularly cereals and soybean meal (FAO, 2006a). We estimate that in 2004 the poultry sector utilized a total of 294 million tonnes of feed, of which approximately 190 million tonnes were cereals, 103 million tonnes soybean meal and 1.6 million tonnes fishmeal.

Estimates put the global use of cereals for feed (all species included) at 666 million tonnes, or about 35 percent of total world cereal use. This implies that in 2004 cereal utilization as feed by the poultry sector represented about 28 percent of the cereal and 75 percent of soybean meal used by the livestock sector.

The estimates for feed utilization by the intensive poultry sector were obtained by applying a two-step approach. The first step estimates total feed use in poultry systems by applying a "utilization approach", i.e. total feed utilization is obtained by multiplying total production (for poultry meat and eggs) by the corresponding feed conversion ratio which reflects both the intensity and efficiency of the livestock system.

The second step involves apportioning the total feed obtained per region based on the concept of "feed baskets". Feed baskets represent the different components that make up a feed ration in any given country. The major elements of feed baskets in intensive poultry systems are usually cereals, oilseeds and fishmeal, while those in mixed systems are to a greater extent made up of agro-industrial by-products (oilmeals, fishmeal) and crop residues, and contain less cereal. In calculating feed use the following assumptions were made:

1. Cereals make up the bulk of the feed baskets in intensive poultry production – an estimated 60 percent. The rest is shared between oilseeds (mainly soybean) and fishmeal (ibid.). However, cereal use for poultry production differs across countries, with maize dominating in Brazil, China and the United States of America, and wheat in the European Union (EU). In mixed systems, we estimate that cereals make up about 30 percent of the feed basket, with the remainder comprising crop residues and agroindustrial by-products.

2. This estimate also assumes homogeneity of poultry production across countries and regions and, therefore, applies an average feed conversion ratio across all regions. For poultry-meat products, an average was taken based on the feed conversion ratios for broilers, turkeys and ducks. For eggs, the feed conversion ratio average was based on the feed conversion ratio for brown-shelled and white-shelled layers. Poultry reared in landless systems are considered efficient users of feed and therefore have lower feed conversion ratios than those in mixed systems.

Demand for feed by the livestock sector has been a trigger for three major global trends: the intensification of feed production, agricultural expansion and erosion of biodiversity. The production of feed has an impact on the environment at various stages of crop production. In terms of the environment, these three trends have had a number of global impacts, which include land and water pollution, air pollution, greenhouse gas emissions, land-use change through deforestation and habitat change, and overexploitation of non-renewable resources.

Environmental Impacts Related to Feed Production

Intensive Agriculture

Intensification of feed production affects land and water resources through pollution caused by the intensive use of mineral fertilizer, pesticides and herbicides to maintain high crop yields. It is estimated that only 30–50 percent of applied nitrogen fertilizer and approximately 45 percent of phosphorus fertilizer is taken up by crops. estimated that about 20 million tonnes of nitrogen fertilizer were used in feed production for the livestock sector. Based on the estimation that the poultry sector utilizes 36 percent of feed concentrates (cereals and soybean), we can attribute about 7.2 million tonnes of nitrogen fertilizer use to feed production for the sector.

Intensive feed production also contributes to air pollution. The application of nitrogen fertilizer to cropland is a major source of air pollution through the volatilization of ammonia. Assuming an average mineral fertilizer ammonia volatilization loss rate of 14 percent, it has been estimated that global livestock production can be considered responsible for a global ammonia volatilization from mineral fertilizer of 3.1 million tonnes of NH3-N (nitrogen in ammonia form) per year. Based on these same assumptions, the poultry sector can be considered responsible for about 1.1 million tonnes of ammonia volatilization from mineral fertilizer per year.

Intrusion into Natural Habitats

Increases in feed production, have to some extent been related to the expansion of cropland dedicated to feed. Feed production to satisfy the feed demand of intensive systems indirectly affects the global land base through changes in land use. Area expansion is in most cases at the expense of forested land (deforestation) cleared for crop production. For example, the land area for soybean production in Brazil increased from 1 million hectares in 1970 to 24 million hectares in 2004 – half of this growth came after 1996, most of it in the Cerrado, with the remainder in the Amazon Basin. According to Brazil's National Institute of Space Research, just over 2.5 million hectares of forest in the Amazon disappeared in 2002, with about 70 percent of the 1.1 million hectare expansion of the agricultural frontier in the Amazon region alone attributed to soybeans. Wassenaar et al. project large hotspots of deforestation in the Brazilian Amazon forest related to the expansion of cropland, mainly for soybean. Changes in land use can have profound impacts on carbon fluxes, leading to increased carbon release and fuelling climate change. In addition to changes in carbon fluxes, deforestation also has an impact on water cycles and increases runoff and consequently soil erosion. WWF estimates that a soy field in the Cerrado loses approximately 8 tonnes of soil per hectare per year.

Erosion of Biodiversity

Feed production is also driving biodiversity erosion through the conversion of natural habitats and the overexploitation of non-renewable resources for feed production.

Intensive feed production contributes to biodiversity loss through land use and land-use change, and modification of natural ecosystems and habitats. The demand for feed has triggered increased production and exports from countries such as Brazil. Between 1994 and 2004, land devoted to soybean production in Latin America more than doubled to 39 million hectares.

The clearing of vast areas of the biologically rich Amazon rainforest and the Cerrado to produce maize and soybeans for feed has led to the loss of plant and animal species.

Overexploitation of Natural Resources

The production of fishmeal for the poultry sector is an important factor contributing to the overexploitation of fisheries. The world's fish stocks are facing serious threats to their biodiversity. The principle source of this pressure is overexploitation, which has affected the size and viability of the fish. FAO estimates that 52 percent of the world's fish stocks are fully exploited, and are therefore producing catches that are already at or very close to their maximum sustainable yield. Current estimates are that around 40 percent of global fishmeal production is used for the livestock sector of which 13 percent is used by the poultry sector.

The expansion of the aquaculture sector and its demand for fishmeal as a feed ingredient has led to a reduction in the use of fishmeal by the poultry sector. Between 1990 and 2006, the share of fishmeal consumed by the poultry industry decreased drastically from about 58 percent to 13 percent. The poultry sector compensated for this loss by increasing the amount of soybean meal used in feed rations. Despite current and projected decline in the sector's use of fishmeal as a feed input, the role of the industrial poultry sector in the overexploitation and depletion of fisheries can not be ignored.

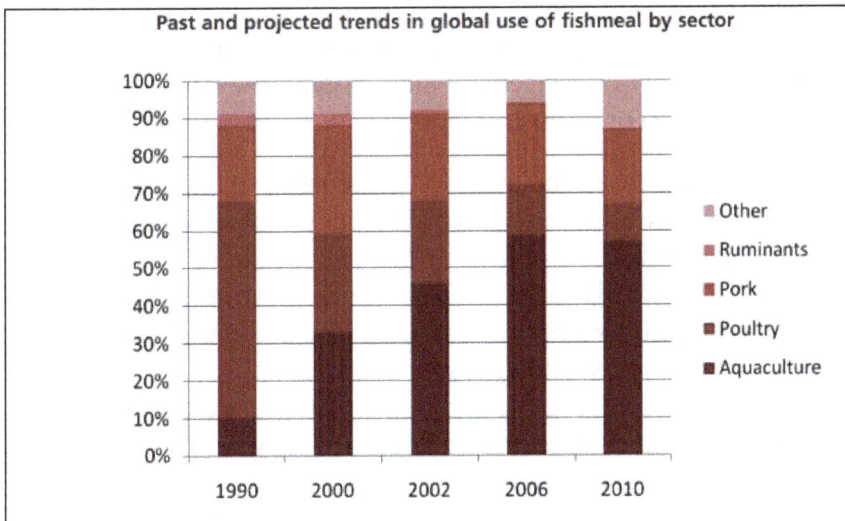

Past and projected trends in global use of fishmeal by sector.

Climate Change

The relatively high energy input in intensive livestock systems has given rise to concerns regarding greenhouse gas emissions and climate change. The energy consumption of industrially produced poultry is relevant because of the production of carbon dioxide (CO_2) along the production chain. Carbon dioxide emissions are produced by the burning of fossil fuels during animal production and slaughter, and transport of processed and refrigerated products, but importantly also through land use and land-use change, and the use of inputs for the production of feed.

On-farm Energy Consumption

On-farm energy consumption includes direct and indirect energy input – direct energy refers to fossil energy used for the production process (e.g. energy input for poultry housing systems), and indirect energy to that used as an integral part of the production process (e.g. feed processing). Due to a lack of information on energy use for processing, this estimation of on-farm fossil fuel consumption is limited to quantifying energy use associated with poultry housing.

The energy used for heating, ventilation and air conditioning systems typically accounts for the largest quantity of energy used in intensive poultry operations. Animal housing facilities are therefore potential sources of carbon dioxide emissions from intensive poultry farms. Other sources of carbon dioxide emissions include energy used for feed preparation, on-farm transport and burning of waste. Generally, on layer farms, artificial heating of housing is not commonly applied, due to the low temperature needs of birds and the high stocking density. The activities that require energy are ventilation, feed distribution, lighting, and egg collection, sorting and preservation. On broiler farms, the main energy consumption is related to local heating, feed distribution and housing ventilation. Quantification of the energy consumption of intensive poultry farms is a complex undertaking because systems are not homogeneous. The amount of energy consumed varies with the technologies applied, the production characteristics of the farms, and climatic conditions.

A rough indication of the fossil fuel related emissions from intensive poultry systems can be obtained by applying the energy requirements, assuming that the energy consumption for heating during the winter in higher latitude countries is equivalent to the high energy use for ventilation in lower latitudes. By applying these estimates to the global total for intensively produced poultry, it is estimated that about 52 million tonnes of carbon dioxide are emitted per year.

Carbon Dioxide Emissions from Slaughtering

Poultry processing facilities use energy to heat water and produce steam for process applications and cleaning, and for the operation of mechanical and electrical equipment, refrigeration and air compressors. In poultry abattoirs, fossil fuel is mainly used for process heat, while electricity is used for the operation of machinery and for refrigeration, ventilation, lighting and the production of compressed air. In an analysis of energy consumption in the EU meat industry found poultry slaughtering to be more energy intensive (3 096 MJ/tonne dress carcass weight) than other meat sectors (1 390 MJ/tonne dress carcass weight for beef and 2 097 MJ/tonne dress carcass weight for pork).

Using the estimates of energy consumption values for poultry, we estimate that carbon dioxide emissions from poultry slaughtering facilities amount to 18 million tonnes. This estimate is obtained by applying the energy consumption data to total poultry meat production and multiplying by the respective carbon dioxide emission factors for both electricity and natural gas.

Carbon Dioxide Emissions from International Trade

International trade in poultry meat contributes significant carbon dioxide emissions – induced by fossil fuel use for the shipping of poultry meat. Estimated carbon dioxide emissions by combining traded volumes with respective distances, vessel capacities and speeds, fuel use of main and auxiliary power generators for refrigeration, and their respective emission factors. Based on this analysis, trade in poultry meat was found to contribute an estimated 256 000 tonnes of carbon dioxide (representing about 51 percent of the total carbon dioxide emissions induced by meat-trade ocean transport). The addition of transportation within national boundaries, involving shorter distances, but much larger quantities and less efficient vehicles, would certainly increase significantly the sector's greenhouse gas emissions related to transportation.

Greenhouse Gases Emissions from Feed Production

Emissions of greenhouse gases such as carbon dioxide and nitrous oxide are influenced in an indirect way by intensification of feed production, which requires energy input for the production of mineral fertilizer and the subsequent use of this fertilizer in the feed production process.

Carbon Dioxide (CO_2)

This greenhouse gas is produced by the burning of fossil fuels during the manufacture of fertilizer. By applying energy use per tonne of nitrogen fertilizer (estimated at 40 GJ per tonne) and the IPCC (Intergovernmental Panel on Climate Change) emission factor for natural gas (17 tonnes of carbon per terajoule) to total nitrogen fertilizer use in the production of feed for poultry production (estimated at 7.2 million tonnes) and applying the ratio of the molecular weight of carbon dioxide to the molecular weight of carbon (44/12) results in an estimated annual carbon dioxide emission of 18 million tonnes – about 44 percent of that ascribed to the livestock sector.

Nitrous Oxide

Nitrous oxide because of the sector's high concentrate-feed requirements and the related emissions from arable land due to the use of nitrogen fertilizer. FAO–IFA reported a 1 percent N_2O-N (nitrogen in nitrous oxide) loss rate from nitrogen mineral fertilizer applied to arable land. By applying this loss rate to the total nitrogen fertilizer attributed to the poultry sector, we estimate that nitrous oxide emissions from poultry feed related fertilizer to be 0.07 million tonnes of N_2O-N per year – about 35 percent of the global nitrous oxide emissions attributed to the livestock sector from mineral fertilizer application.

Overall, intensive poultry production (indirectly and directly) contributes an estimated 3 percent of the total anthropogenic greenhouse gas and is responsible for about 2 percent of the total greenhouse gas emissions from the livestock sector. This estimate however does not include emissions from land use and land-use change associated with feed production or emissions related to transport of feed.

Technical Mitigation Options

The magnitude of environmental impacts is highly dependent on production practices and especially on manure management practices. Lack of awareness and capital are often cited as the two factors hampering the implementation of such practices.

Farm Management

Taking environmental issues into account in all management strategies at the farm level can reduce the impacts felt at the level of production.

Odour emissions can be controlled by:

- Minimizing the surface of manure in contact with air – frequent collection of litter (once a week in dry seasons and twice a week in rainy seasons), closed storage (bags or closed sheds);

- Cooling animal manure, achieved as a positive side effect of cooling the animal houses – cooling systems can be equipped with biofilters and air scrubbers that trap odours from the ventilation airflow;

- Lowering litter's water content – achieved by the incorporation of hydrophilic products such as hashes, rice husk, peanut husk, dust or sawdust;

- Applying deodorant products to feed or directly to animal houses; and

- Building wind protection structures.

The proliferation of flies and mosquitoes can be controlled by:

- Minimizing the surface of manure in contact with air – frequent collection of litter (once a week in dry seasons and twice in rainy seasons, i.e. at shorter intervals than the length of the larvae development cycle), closed storage (bags or closed sheds);

- Lowering litter's water content – achieved by the incorporation of hydrophilic prod ucts such as hashes, rice husks, peanut husks, dust, sawdust or available dry crop residues;

- Applying insecticides (this practice may however have significant public health-related side effects);

- Building wind protection structures;

- Positioning nets around the farm.

Rat proliferation can be controlled by:

- Minimizing feed losses during storage and feeding;

- Raising cats or keeping snakes in cages close to the poultry barn to scare rats; and

- Use of poison or traps.

Visual impact and landscaping can be improved by:

- Use of screening trees around the farm facility to reduce the visual impact of farm infrastructure and of noise, dust, light and odour;

- Use of the natural topography and terrain of the site and the existing vegetative cover to maximize visual screening; and

- Use of construction materials that minimize visual impact.

Animal Waste Management

Soil and water pollution is controlled through the implementation of good fertilization practices. In brief: environmental risks are reduced when manure is applied in amounts and at times that correspond to crop or fish-pond uptake. Water pollution is often an acute problem in waterfowl production, especially when the flock is concentrated on relatively small ponds. There is currently a lack of information with regard to the effects of waterfowl production on surface water and groundwater resources.

Water- and food-borne disease propagation can be prevented by:

- Storing manure in closed buildings or bags – a storage system allows producers to hold manure until a convenient and optimum time for use; storing poultry manure in closed buildings reduces the emissions of gaseous compounds to the air, and the risk of environmental contamination as compared to the risk associated with leaving manure exposed;

- Storing the manure for one to two months before its application on land or fish ponds;

- Composting manure – potentially reduces or even eliminates certain pathogens and fly larvae, and improves the handling characteristics of manure and other residues by reducing their volume, weight and moisture content (most manure and other organic residues usually contain high nitrogen content and are, therefore, subject to nitrogen loss during composting);

- Drying (with machine or by spreading out) – minimizes the moisture content of manure, inhibits chemical reactions, and thus reduces emissions (the best way to prevent ammonia emissions from poultry litter and manure is to reduce microbial decomposition, which can be accomplished by drying the freshly produced manure as soon as possible and keeping it dry);

- Timing and rate of manure application – this is a critical management factor; manure must be applied at the correct time of year to prevent losses to surface water, groundwater and the atmosphere, and to optimize the utilization of manure nutrients by growing plants; proper timing is a function of several variables, including weather, soil conditions and stage of crop growth; and

- Dead-bird management and disposal, which must comply with legally accepted practices including rendering, composing, incineration and burial; a contingency plan should be in place for disposal of large numbers of dead birds in the event of disease outbreaks; in addition, consideration should be given to impacts on the physical environment – e.g. burial pits should be at least 3 metres above the maximum groundwater table.

Nutrition Management

Nutritional management aims to reduce pollution load by limiting excess nutrient intake and/or improving the nutrient utilization efficacy of the animal. It not only affects the quantity of mineral outputs from animals and the characteristics of manure, but also has cross-media effects – reducing the pollution load of soil, water and air. Nutrition management can also allow improvement to feed conversion ratios through optimal diet balancing and feeding regimes, and improvement to feed digestibility. This reduces the amount of feed used per unit of livestock product. Relevant measures include:

- Formulating feeds that closely match the nutritional requirements of birds in their different production and growth stages to reduce the amount of nutrients excreted; options in this category include phase feeding, split-sex feeding or feed formulation on an available-nutrient basis;

- Use of low-protein diets supplemented with amino acids, and low-phosphorus diets with highly digestible inorganic phosphates;

- Improving feed digestibility and nutrient bioavailability through the use of dietary supplementary enzymes such as phytase, highly digestible genetically modified feedstuffs such as low-phytate maize, and highly digestible synthetic amino acids and trace minerals; and

- Using good quality, uncontaminated feed (e.g. in which concentrations of pesticides and dioxins are known and do not exceed acceptable levels) which contains no more copper, zinc, and other additives than is necessary for animal health.

Feed Production

The key to reducing the negative environmental impacts associated with intensive agriculture for feed production lies in increasing efficiency, i.e. increasing production while reducing the use of inputs that adversely affect the environment. The negative effects of feed production can be greatly reduced with appropriate cultivation (e.g. minimum tillage), integrated pest management (IPM) and targeted fertilizer inputs. Technologies are available for many different environments to conserve soil and water resources and to minimize the use and impact of inorganic fertilizers and pesticides.

Good agricultural practices require the application of available knowledge to the utilization of the natural resource base in a sustainable way for the production of safe and healthy food. Management of resources such as soil and water by minimizing losses of soil, nutrients and agrochemicals through erosion, runoff and leaching into surface water or groundwater is a criterion for good agricultural practice. Good agricultural practice will maintain or improve soil organic matter through the use of appropriate mechanical and conservation tillage practices; will use soil cover to minimize erosion loss by wind or water; and will ensure that agrochemicals and organic and inorganic fertilizers are applied in amounts, at times and using methods, appropriate to agronomic and environmental requirements.

IPM uses an understanding of the life cycle of pests and their interactions with the environment, in combination with available pest control methods, to keep pests at a level that is within an acceptable threshold in terms of economic impact, while giving rise to minimum adverse environmental and human health effects. Recommended IPM approaches include: use of biological controls such as predators, parasites and pathogens to control pests; use of pest-resistant varieties; mechanical and biological controls; and, as a last resort, chemical controls including synthetic and botanical pesticides. Other IPM approaches encompass pesticide application techniques that aim to increase the efficiency of chemical applications.

Minimal tillage practices in agronomic crops such as soybean and maize reduce the loss of nutrients from the field, they also improve the water-stability of soils and reduce soil erosion; this often results in higher levels of soil organic carbon and reduces carbon emissions.

Enhancing the efficiency of water use in feed production by improving irrigation efficiency and water productivity is a further method of reducing adverse environmental impacts. Water productivity can be improved by methods including selection of appropriate crops and cultivars, better planting methods, minimum tillage, timely irrigation that matches water application with the most sensitive growing periods, nutrient management and drip irrigation.

Most of the issues associated with poultry production, as environmental impacts related to backyard or mixed extensive systems are marginal because of the limited concentration of wastes and reliance on locally available sources of feed, such as food residues, crop residues or feed collected by free-ranging birds. The review has also demonstrated the need to look beyond the farm level in order to understand the sector's impacts on the environment, as many of the impacts of production are felt beyond the point of production.

Generally, the environmental impacts of the sector are substantial. Poultry production is associated with a variety of pollutants, including oxygen-demanding substances, ammonia, solids, nutrients (specifically nitrogen and phosphorus), pathogens, trace elements, antibiotics, pesticides, hormones, and odour and other airborne emissions. These pollutants have been shown to produce impacts across multiple media. These impacts can be summarized as follows:

- Surface water impacts: Impacts are associated with waste spills, as well as surface runoff and subsurface flow. The oxygen demand and ammonia content of the waste can result in fish kills and reduced biodiversity. Nutrients contribute to eutrophication and associated blooms of toxic algae and other toxic micro-organisms. Human and animal health impacts are associated with drinking contaminated water (pathogens and nitrates) and contact with contaminated water (pathogens and Pfiesteria). Trace elements (e.g. arsenic, copper, selenium and zinc) may also present human health and ecological risks. Antibiotics, pesticides and hormones may have low-level but long-term ecosystem effects.

- Groundwater impacts: Impacts associated with pathogens and nitrates in drinking water may cause underlying groundwater to become unsuitable for human consumption.

- Air/atmosphere impacts: Impacts include those on human health (caused by ammonia, hydrogen sulfide, other odour-causing compounds, and particulates), and contribution to global warming (due to carbon dioxide and nitrous oxide emissions from the production process and other related activities such as feed production and transport of finished products). Additionally, volatilized ammonia can be re-deposited and contribute to eutrophication, acidification and damage to vegetation and sensitive ecosystems.

- Soil impacts: Nutrients and trace elements in animal manure can accumulate in the soil and become toxic to plants.

Other indirect impacts include ecosystem destruction and biodiversity erosion associated with the expansion of feedcrop production into natural habitats and the overexploitation of non-renewable resources for feed production.

Compared to other livestock species, however, poultry performs well from an environmental perspective. A substantial comparative advantage that poultry has over other animal sectors relates to its efficiency in feed conversion. Poultry's feed conversion ratio represents a major contribution to the profitability of the industry in terms of reduced feed inputs as well as in waste output. For cattle in feedlots, it takes roughly 7 kg of grain to produce a 1 kg gain in live weight. For pork, the figure is close to 4 kg per kg of weight gain, for poultry it is just over 2 kg, and for herbivorous species of farmed fish, such as carp, tilapia, and catfish, it is less than 2 kg. Another comparative advantage lies in the low water content and high nutrient content of poultry manure. It is, thus, often handled with more care than manure from other species – especially pigs – as recycling is generally economically profitable.

Technologies exist that have the potential to produce substantial reductions in environmental impacts. The problem is one of cost, corresponding incentives/disincentives and awareness. Given the strong reactivity of the sector (large companies, foreign direct investment, demand growth), getting economic incentives and disincentives right within a framework of market forces should be sufficient to minimize environmental impacts.

References

- Diseases-causes, maintaining-healthy-flock, diseases-health-management, poultry, livestock, agriculture, farms-fishing-forestry, industries: business.qld.gov.au, Retrieved 7 June, 2019

- "California modifies virulent Newcastle disease quarantine boundaries". Feedstuffs. 27 February 2019. Retrieved 6 April2019

- Overview-of-erysipelas-in-poultry, erysipelas, poultry: msdvetmanual.com, Retrieved 8 July, 2019

- Robyn Alders; Spradbrow, Peter (2001). Controlling Newcastle disease in village chickens : a field manual. Canberra: ACIAR. ISBN 978-1863203074

- Types-of-disease, disease, health: poultryhub.org, Retrieved 9 August 2019

- Mites-of-poultry-v3343223, ectoparasites, poultry, msdvetmanual.com: Retrieved 10 January, 2019

- "Leaky Vaccines Enhance Spread of Deadlier Chicken Viruses". 2015-07-27. Retrieved 2018-11-03

- Haemoproteus-infection-in-poultry, bloodborne-organisms, poultry: msdvetmanual.com, Retrieved 11 February, 2019

- Rossi, Gary D. Butcher and Fred (2015-06-19). "Prevention and Control of Fowl Pox in Backyard Chicken Flocks". Edis.ifas.ufl.edu. Retrieved 2016-03-02

- Health-management, health: poultryhub.org, Retrieved 12 March, 2019

Permissions

We would like to thank the editorial team for lending their expertise to make the book truly unique. They have played a crucial role in the development of this book. Without their invaluable contributions this book wouldn't have been possible. They have made vital efforts to compile up to date information on the varied aspects of this subject to make this book a valuable addition to the collection of many professionals and students.

This book was conceptualized with the vision of imparting up-to-date and integrated information in this field. To ensure the same, a matchless editorial board was set up. Every individual on the board went through rigorous rounds of assessment to prove their worth. After which they invested a large part of their time researching and compiling the most relevant data for our readers.

The editorial board has been involved in producing this book since its inception. They have spent rigorous hours researching and exploring the diverse topics which have resulted in the successful publishing of this book. They have passed on their knowledge of decades through this book. To expedite this challenging task, the publisher supported the team at every step. A small team of assistant editors was also appointed to further simplify the editing procedure and attain best results for the readers.

Apart from the editorial board, the designing team has also invested a significant amount of their time in understanding the subject and creating the most relevant covers. They scrutinized every image to scout for the most suitable representation of the subject and create an appropriate cover for the book.

The publishing team has been an ardent support to the editorial, designing and production team. Their endless efforts to recruit the best for this project, has resulted in the accomplishment of this book. They are a veteran in the field of academics and their pool of knowledge is as vast as their experience in printing. Their expertise and guidance has proved useful at every step. Their uncompromising quality standards have made this book an exceptional effort. Their encouragement from time to time has been an inspiration for everyone.

The publisher and the editorial board hope that this book will prove to be a valuable piece of knowledge for students, practitioners and scholars across the globe.

Index

www.ingramcontent.com/pod-product-compliance
Lightning Source LLC
Chambersburg PA
CBHW061945190326

41458CB00009B/2787